Financial Planning For The Small Building Contractor

Derek Miles

Intermediate Technology Publications

Published by Intermediate Technology Publications Ltd., 9 King Street,
London WC2E 8HN, U.K.

ISBN 0 903031 55 8

Printed by the Russell Press Ltd., Gamble Street, Nottingham NG7 4ET
Telephone: (0602) 74505

Contents

Acknowledgements

The material upon which this volume is based was prepared for an I.L.O. African Regional Construction Management Course sponsored by N.O.R.A.D. (the Norwegian Aid Agency). The Intermediate Technology Development Group gratefully acknowledges their permission to publish, as well as the assistance of the Ministry of Overseas Development in financing earlier work in this field, and the publication of this book.

Preface

Management is about getting things done. Construction management is about getting things built. This book is the result of a decade of experience in helping small contractors in developing countries establish themselves and to run their businesses effectively. I hope that it will help – both as a text book and as a basis for training courses – the managers and owners of small contracting businesses to improve their managerial capacity. This should in turn help them to become better employers, by offering more permanent jobs, as well as helping their clients by making them more responsive to the needs of their customers.

As designs, materials and components have become more complex, and demand has risen, the construction industry has taken on a key role and its performance impinges on all other sectors of the national economy. Thus construction costs are not merely a matter of concern to the clients of the industry, but must also concern the nation and its government. Indeed, construction is often responsible for creating more than half a nation's wealth in terms of fixed assets, so value for money must be a prime concern.

Speed of construction as well as cost is important. Unfortunately local construction industries are often criticised by economists and others for failure to meet completion dates and, unfortunately, these criticisms are often justified due to inadequate performance. Indeed, there is a vital link between time and money in construction management, and contractors usually find that quicker jobs lead to lower costs and bigger profits.

But governments and international aid agencies have a role in helping their local construction industry to become more competitive. One of the earliest efforts in this field was an Intermediate Technology Development Group project for 'Technological and procedural guidance to the construction industries of less developed countries', which was financed for the initial period 1969-72 by the British Ministry of Overseas Development. During this period ITDG co-operated on the development of training material with the Kenya National Construction Corporation Ltd, which was started as a joint venture between the Kenya Government and NORAD (the Norwegian aid agency) in 1967.

More recently, the International Labour Office, with financial support from NORAD, has set up a project to promote training in practical construction management within the African region. The immediate objective was defined as: 'to create in the participating countries a basic capability for delivering management training to small-scale contractors', while the longer term objective is to improve the overall managerial and economic performance of the contractors.

The material upon which this book is based has been developed over the period. The approach is decidedly practical, with emphasis on providing ideas and techniques which the reader can apply in a straightforward way to increase his knowledge of, and control over, his business. Most of these ideas and techniques are just as relevant to good management in the public sector direct works agency as to running a private business for profit. Saving time — and saving money — are the twin themes.

Please note that the figures used in the examples and exercises are intended only for working out the exercises, and do not reflect the actual costs of purchasing plant and machinery, hiring equipment, etc.

No book of this kind can come from the knowledge of one individual alone. The author willingly acknowledges that he has drawn on ideas and experience from many people and many sources over the years. I would, however, like to mention four people with whom I have had the pleasure of talking, and working as co-lecturer, on a number of occasions — and in a number of countries. They are Dr Colin Guthrie and Mr John Andrews of the ILO, and Mr Folkward Vevstad and Mr Jostein Fjellestad, consultants to NORAD. Errors and omissions are, of course, mv own. Finally, mv thanks to the editorial staff of Intermediate Technology Publications for their care and effort in preparation.

Derek Miles

Chapter One
Aims and Objectives

Growth and purpose. Planning for growth. Written objectives. Aiming at a fair profit. Other objectives. Advantages and dangers of the family business. Looking at technology – application of labour-intensive technologies. Formulating objectives. Defining a development path. Taking account of the current situation.

Where are we going?

There is not much point in getting ready for a journey if you have no clear idea of where you want to go. In the same way most people would agree that there is not much point in thinking about starting up some form of business enterprise if you have no clear idea of how you want it to progress and what you hope it will achieve.

Growth and purpose

Before the prospective building contractor starts to think about the ***growth*** of his business into a prosperous enterprise, he should first give some thought to his ***purposes*** and intentions in entering this highly competitive field. Some people are attracted into the industry because they see existing contractors with large cars and who appear to be doing well, so they think that it is an easy way to make money. Such a person might think that once he has decided to become a building contractor, all he has to do is register a suitable trade name and he will turn into one overnight. He would be quite wrong!

Muddling along

Of course we know of people who muddle along in the industry getting one or two small jobs a year and filling in their time with farming or keeping a shop. These people cannot properly be described as real builders, and they certainly cannot be described as professional builders. They are of little use to clients, since they lack real experience and, although they are occasionally awarded contracts when a very low tender is submitted , they cost more in the long run because they need greater supervision. They are of little use to their employees, who are likely to lose their jobs as the contract ends. Finally they are of little use to themselves, since their earnings from the building industry give them no encouragement to progress.

Four roads

In any country there are four possible roads that a business might take:

1. Failure — due perhaps to cut-throat competition but very often to poor management and lack of foresight;
2. Muddling along — with the proprietor of the business lacking ambition and initiative, he lives from day to day as his business totters from crisis to crisis but just about keeps going;
3. Unplanned expansion — taking advantage of luck or an expanding national economy, a firm may well show apparently impressive progress. It will get bigger if more jobs turn up, but the danger is that there is no clear plan so that if times become less favourable the businessman just waits and hopes for things to get better;
4. By following a ***business policy*** based on appropriate ***aims*** and ***objectives*** build a reputation which will ensure that clients will award contracts and the business will continue to grow even when work is scarce.

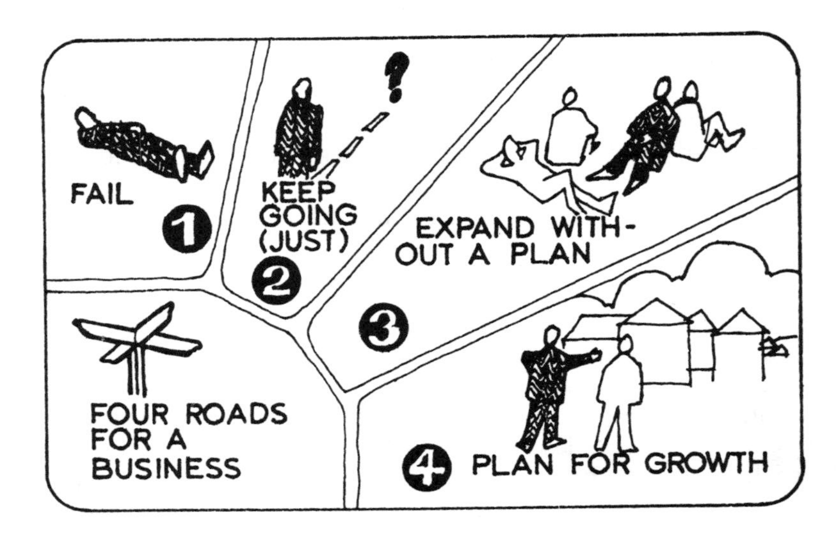

Route number 4

No businessman can be certain that his business will prosper and grow and, since we do not have the gift of foretelling the future, no-one can be sure that any particular business is headed along road 1, 2, 3 or 4. The job of a manager would be much easier if he did know this, but no consultant or expert can give him any such guarantee. Unfortunately this book cannot lead to the certain conclusion of success, although I hope that it will suggest ways in which a thoughtful and hard-working businessman can improve his chances. He could well start by recognising the sense of attempting to follow Route No.4 and giving careful thought to why he ***really*** set up in business in the first place.

Written objectives

The first requirement for anyone who hopes to become a successful businessman is to ensure that his objectives (what he is in business to do) are clearly thought out and ***set down on paper.***

A major step

Setting up a business is a major step for anyone, since it means committing oneself to spend a great deal of money, time, effort and energy to provide a worthwhile and acceptable service to attract clients and customers. Nobody would do this without a reason, and the reason may be quite obvious and simple such as 'to make a profit and improve my standard of living'. Even so, it is still worth writing the objectives down, so that they will be available after the business has been operating for a while to see whether it is working out according to plan.

Checking performance

It is very easy for the businessman (who is always busy and never has as much time as he would wish) to lose sight of his original objectives. Thus he should regularly check his actual performance against stated objectives, so that he can find out where this performance is less effective than he intended. He can then take steps to change his operations in order to achieve greater success.

Some possible objectives

Some possible objectives which might lead someone to start a business and keep it in operation are:

1. To stay in business

2. To make a profit
3. Growth
4. Provide a service
5. Jobs for the family
6. To provide employment
7. To be an efficient organisation.

Profit

Many businessmen would simply say that their reason for being in business is simply to 'make a profit'. Others would not even aim as high as that, and are content merely to survive or 'stay in business'. But most people want to do both. In fact to stay in business without making a profit is very difficult and, unless you start off very rich, it is probably impossible. Providing the profit is honestly earned, the businessman has no reason to be ashamed of it, and the contractor who submits a tender in which the profit margin is carefully hidden really fools nobody. Profit is not only an incentive to the businessman, it is his safety margin if things go wrong. Thus the client should not be envious if his contractor earns a reasonable profit on his work, because profit is also a safety margin for the client in that it guarantees that the contractor will find it worthwhile to get the job completed on time and in accordance with the specified quality standards.

Profit is essential

Profit therefore is an important (possibly even the most important) business objective. Without it a company will have nothing to plough back into the business to expand its activities nor money to pay interest on or repay loans and overdrafts or pay dividends to its shareholders.

Growth

Although staying in business and making a profit are usually basic objectives for a building contractor, they do not tell the whole story. Even when their businesses have reached a stage of development where they provide a comfortable living for the proprietor and his family, many businessmen enjoy the feeling that their firm is growing and that it is able to carry out more and more work as the years pass.

Good for the community

Growth need not just be a selfish objective leading to personal riches and success. If that were the only pay-off from growth, it would not be worth while for governments and international agencies to help to develop and encourage the growth of indigenous contracting industries. The reason for these training and development activities is that governments realise and understand that a country must first have stable, efficient building firms and units if there is to be a hope of development plans being fulfilled. These plans are geared to the increasing aspirations of people and, whether they seek education, better health facilities, housing or jobs, it is almost certain that buildings will be required to meet their needs. A competent building contractor is an asset to the country's economy and to its society, as well as to his own family and friends.

Too much profit

It may be said that, if growth and profit are a good thing, surely it is not possible to have too much profit. Certainly a business will fail if no profit emerges, but the businessman should not be greedy and attempt to make too much. For example, it might be possible for a contractor with a name for careful and high quality work to increase his profit in the short term by taking short cuts and lowering standards. However, this would be very short-sighted.

Reputation

Action of this kind would mean that his chances of future work would be much less once his poor work became evident to the client. Even if he managed to get away without expensive remedial works, his reputation would have suffered permanently. Reputations are hard-won over the years, and they can be lost overnight.

Fair profit

So it is foolish to have as the only aim of the company the making of the highest possible immediate profit. A much better objective for the serious businessman is to make a ***fair profit*** consistently over the years.

Other aims

The aims of the businessman will result in his adopting a suitable organisation and style of running his firm. Indeed most businesses have more than one specific aim, depending on the personalities and interests of their owners. One possible objective may be rapid ***growth*** of turnover. A young keen manager may intend to build up, over a period of 10 or 20 years, an organisation capable of tackling large civil engineering projects such as roads, bridges and dams. He may have no particular interest in making large profits, but simply wish to build a worthwhile business that he can be proud of and hand on to his children.

More investment

If this is the case, he will be very keen to employ graduates and other skilled men, and to buy modern plant and equipment. These may not be justified by his present contracts, but would fit in with his plans for the future. Thus his approach to staffing his business and planning capital investment in plant and equipment would be fundamentally different from that of another businessman who was content to stabilise his turnover at the current level. Although their approaches are different, neither can be labelled as right or wrong by any outside norm or standard. For a growing business, a higher level of investment is right simply because it accords with that particular firm's individual objectives.

Some avoid growth

But not every company wants to grow. Many small businessmen are happy as they are. They know all their employees, customers and clients by name. They are able to keep their knowledge of the activities of their businesses in their heads, possibly aided by simple records in a few small notebooks. And perhaps they are realistic enough to have an uncomfortable feeling that they might not be able to cope with running a much larger organisation.

Realism pays

This may not be wrong. It would be foolish to encourage every business to expand, regardless of its individual characteristics. Where the alternative would be over-expansion leading to failure and bankruptcy, it is obviously better to be realistic and keep things going as they are. In these circumstances, the objective would not be growth of turnover but rather the maintenance of a steady order book from one year to another. Such an objective would show up in day to day decisions. For instance, the number of staff would not be increased but simply replaced as they leave and, if specialised tools or plant were needed to carry out a contract, they might be hired instead of bought outright.

Self respect

Although profit and business growth may be important objectives, it is clear that these are not the only reasons for a decision to be involved in business activity. If a man is a trained engineer or a skilled craftsman, he may feel that this is a way of repaying the time and dedication that went into providing that training. In this case, he will probably be quite happy so long as the business gives him the self respect and satisfaction that comes from exercising personal skills providing, of course, that it generates sufficient profit to provide a stable income for him and his family.

Providing a service

One aspect of self respect in a businessman is to take a genuine pride in providing a service to his neighbours and to the community at large. Buildings are the starting point for the provision of many community services, and contractors are proud to know that their fellow-citizens will meet in a hall which they have constructed, or that their children will be taught in a school which they have built.

Jobs for the family

Another good reason for wanting to run a successful business is that it gives a chance to provide jobs. In a country where there is a great deal of unemployment and good jobs are hard to find, a man who builds a successful business and is able to offer good jobs can be a real benefactor. In some cases these jobs may be filled by members of one's own family and, providing they are capable of doing the job and are willing to work hard and carry out instructions, this can be a good idea.

Good for organisation?

This may be good for the family, but it is not necessarily always good for the organisation. A good reason for a contractor to bring his family into his business is because he feels that he can trust them more than strangers. A good family business can be one of the most stable, efficient and trustworthy organisations possible, but there must be a clear understanding from the start that salaries, wages and promotion depend on hard work and merit, and that equal opportunities are available to non-family members of staff if they show that they deserve them.

Dangers

But a policy of giving preference to employing family and friends can also have its dangers. If a stranger lets you down, you can quite easily reprimand him and warn him that if anything else goes wrong he will be dismissed. But with relations things are more complicated. They may see themselves as specially privileged and rely on family feelings to make them immune from criticism. In turn, this can lead to jealousy and bad feeling among other employees.

Duty is mutual

It is understandable and right that a successful businessman should wish to share his personal prosperity with less fortunate members of his family, and he may even believe that he has a direct duty to help them. But they should also feel a duty. It is their duty to repay his thoughtfulness in offering them a job by working hard and setting an example to non-family employees. If you do employ members of your family in your firm, it is vital that they should understand quite clearly from the start that they will be judged on the basis of their performance. The contractor can help them by making a special effort to provide suitable training in order to build up the skills and knowledge that will make them good employees.

Increasing employment opportunities

Once the family has been looked after, most public-spirited people would hope to be able to help others in their own village, town or local community. Thus there are other community-minded employers who deliberately increase the size of their businesses so that they can offer better employment opportunities to more of the men in their area. It is bad that large numbers of men should be out of work. It is bad for their individual morale as well as a serious waste of their country's resources. So it is always sensible for a contractor to think about the technology which he intends to use to carry out each operation on every contract that he is awarded.

Looking at technology

Some people think that using manpower, animal-drawn equipment or deliberately simple machinery is old-fashioned, while users of expensive modern equipment, such as crawler tractors and powered scrapers, show themselves as modern and efficient contractors. This is far too simple a viewpoint, as the test of a technology should be whether it does the job best in the particular conditions, climate, physical, social and economic environment which prevail locally. In many developing countries, where there are high levels of unemployment, wages are relatively low, many of the more sophisticated items of plant and machinery have to be imported, and fuel, repairs and replacements are extremely expensive. In these circumstances, many of the more progressive construction units are deliberately aiming to maximise the use of simple labour-intensive methods and trying to avoid complex capital-intensive techniques. They find that this objective is not only justified in social terms, but reduces fixed capital requirements and improves cash flow, as well as lowering overall costs over a period.

Labour-intensive technology

These forms of intermediate or labour-intensive technology have been shown to work in many different parts of the world, but they only work if the labour itself is properly productive. It may be worthwhile to employ 100 men with simple tools rather than ten with expensive machinery because the extra wages will be more than paid for by the saving in investment and running costs. In fact, labour-intensive technology is far from old-fashioned, and more and more countries are trying out ways of applying it to improve employment levels and save foreign exchange.

Labour must be productive

Although the technology upon which labour-intensive techniques are based is certainly simpler, it increases the responsibility of management to ensure that the additional labour which is employed is sufficiently productive. Labour-intensive techniques do not mean employing two people to do a job that could be done perfectly well by one, and employees must be told clearly what they have to do and how long they have to complete it. If manual methods are chosen, supervision by the contractor and his foreman has to be improved.

Checking costs

It is vital that the contractor should set up a system for checking costs as well as the benefits of using a lot of labour and little machinery. Otherwise unit costs may rise, leading to higher tender figures on future contracts and falling turnover. Thus loose supervision could lead eventually to a loss of business and men being laid off in the end: just the opposite of what the contractor intended.

An efficient organisation

A further good reason for a contractor to remain in business after he has provided for his own and his family's immediate financial needs is the satisfaction to be gained from running an efficient organisation. Whether one is running a business, delivering a lecture or playing a game, there is a great deal of satisfaction to be gained from doing a job well. The best businessmen enjoy their business life, and are always looking at ways in which costs could be saved, technical problems could be solved or output expanded.

Recognise different objectives

The foregoing has been intended to show that there are more reasons for being in business than making as big a profit as possible as quickly as possible. Different people have different aims and different ambitions. That does not matter and it is not possible to lay down a set of priorities that every businessman would accept. What is important is that every businessman should understand his own personality and recognise the real reasons for his commitment to the construction industry. Only then can he set out logically to achieve them.

Aims

Once the businessman has recognised his own concerns and priorities, he can go on to amalgamate them into a general overall aim. This should be specially tailored to suit the level of development of the community, the chosen nature of the business, and the personal feelings of the contractor or businessman. To have a general aim, however, is not enough. The manager must go on to work out ***specific tactical objectives*** by which this aim can be achieved.

Objectives defined by the manager

The definition of business objectives is a job for the top manager himself, although they should also be discussed with those employees who will be expected to play a part in achieving them. Vague hopes are not objectives, because they will be of no value in decision-making or planning and there is no way of telling at the end of an operating period whether or not targets have been achieved. In order to be of value in running the business, objectives should be:

1. Consistent with overall aims
2. Measurable and specific
3. Realistic in terms of resources
4. Acceptable to employees
5. Good for the community.

OBJECTIVES should be:

- CONSISTENT

- MEASURABLE
- REALISTIC

- ACCEPTABLE
- GOOD FOR COMMUNITY

Consistent

It is obvious that individual objectives should be consistent with the overall expressed aim of the business, and should normally play some part in achieving this. Also all objectives should be mutually consistent.

Measurable

Objectives (as distinct from the general aim) should be quite specific and measurable. If this is to be the case, all objectives must be precisely described in terms of:

1. Quantity
2. Quality
3. Cost
4. Time.

An example might be 'ten per cent growth of turnover on building works next year' or 'a fifteen per cent increase in output of concrete blocks with no increase in wastage and a ten per cent reduction in unit cost'. It is most important that output figures or cash costs should be stated so that a check can be made at the end of the year to see whether the objectives have been achieved.

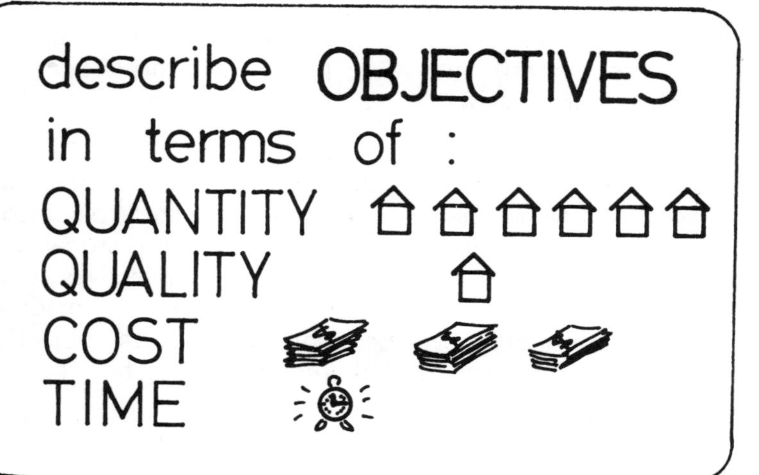

Realistic

Objectives must be realistic in terms of available resources, particularly financial resources. It is important for all businessmen to understand the need for forecasts of cash flow as well as profit to ensure that there is always enough money in the bank to pay wages and bills promptly, but this applies with even more force in building and civil engineering where individual contracts represent large sums of money. Specialist businesses must also pay due regard to the implications in terms of manpower resources, since recruiting or training new staff and operatives will take time. At the same time objectives should be real targets and, although they must be capable of being met, it would be pointless to set objectives which could be easily achieved without any real effort.

Acceptable

If objectives are to be achieved, they must be the objectives of everyone working in the organisation rather then just those of the manager himself. Thus they must be acceptable to all those involved and preferably worked out with them. All employees must know what is expected of them, and they are likely to perform better if their performance will be judged against clear targets defined in advance.

Good for community

The firm's objectives should be in a form that can be appreciated as advancing the well-being of the community at large as well as the private interests of the owner of the firm. This is important because employees are citizens first and employees second, and most people will not co-operate wholeheartedly in selfish and anti-social objectives.

Manager's job

The manager must remember that the setting of objectives is the one activity that can never be delegated to someone else. It is the one thing that no employee can reasonably be expected to do for him. The setting of objectives is the supreme test of the wisdom and foresight of a manager, and time spent on it will be amply repaid as the objectives are striven for and achieved, thereby benefiting both his firm and his community.

Clarity

It may seem obvious to say that objectives should be clear and unambiguous, but it is surprising how often a manager has an unwritten objective at the back of his mind but fails to explain it to his employees. So, even with the best of intentions, his staff have to guess at what he really wants and may find themselves conscientiously working towards some other goal altogether. This criticism applies to everyday instructions as well as annual objectives. More damage is done in industry by muddle and lack of understanding than could be achieved by a gang of vandals bent on sabotage. One good way of checking whether your intentions have been properly understood is to follow up by asking the employee a question related to the instruction. Of course the usual follow up "do you understand?" is worse than useless, since no-one will ever answer "no" and appear foolish.

Payment by results

Once a system of clear objectives has been set up, staff should be rewarded in accordance with their performance in achieving targets agreed by themselves in advance. Most construction managers realise that a man who has to dig a ditch or a foundation trench will perform best if he is told clearly what he has to do and how long he has to do it in. But many fail to apply the same lesson to non-manual employees. They should also have their objectives, so that they know what is expected of them. When employees consistently reach or beat objectives, their salaries should be increased accordingly. When others consistently fail, the reasons for this must be carefully investigated. Over a period, a system of monitored objectives coupled with rewards according to results should put the organisation well on the road to running itself, leaving the management more time to examine possibilities for expansion into new markets.

Development path

The manager who has really worked out and recognised his company's objectives and the aim towards which they lead may be said to have defined a ***development path*** for his organisation. At regular periods, perhaps every month or every quarter, he can check actual achievement against the stated objectives to see how well he is doing. He can also refer to them when circumstances change and new decisions have to be made, so as to take account of the new factors which may modify his approach to the aim of steady progress.

Modifying objectives

Examples of the factors which might lead a manager to modify his objectives are general market factors, such as a reduction in the availability of building work leading to severe competition, or general economic factors, such as a credit squeeze resulting in a reduction in bank overdraft facilities. In addition the contractor's personal circumstances might change, leading to a change in the way in which he views his commitment to his business.

Current situation

All of these factors which are liable to modify a firm's basic objectives can be described as the ***current situation*** facing that business. They are the realities that surround business life and which gradually become apparent during the normal course of business activity.

Business policy

Thus there is one additional stage in overall business planning after objectives have been formulated. This is the formulation of business policy, which can be expressed as a simple equation:

development path + current situation = business policy

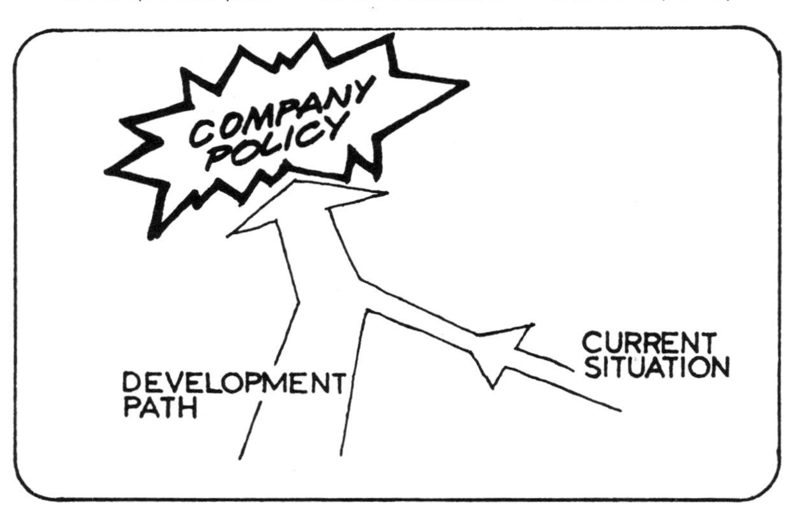

Regular review

The policy of any organisation should be regularly reviewed to take account of changes in its operating environment, including changes in the position of customers and the supplies of services and resources. However, once the development path has been clearly thought out and expressed, it should stand until such time as there is a need to take a fresh look at business objectives.

Policy is flexible

Thus, although the directors of a business must have clear ideas of the path to development that they wish to follow, their actual policies will be kept flexible and will take account of real life problems as they arise. Every year, or more often if there is a sudden change in the business outlook, they will review plans and programmes to take account of past performance and current prospects.

Communications policy

Once a policy has been formulated, the next stage is to ensure that it is properly communicated to all the employees who will be involved in its implementation. If they are to work well and effectively, they must know what they should be doing. But, in addition, they are much more likely to work thoughtfully towards reaching policy goals if they understand *why* the policy has been set.

Participation the best policy

Old fashioned management might disagree and say "I pay staff to do as they are told". But research shows that participation is the best policy, since it leads to better morale and therefore to a more loyal work force.

Intermediate management

This applies with particular force to the 'intermediate management' of an organisation, such as the site foremen and charge hands. They are responsible for the day-to-day and hour-to-hour interpretation of policy on the building site, and are the front-line ambassadors for the owners and senior management. If they are to have real confidence in putting this policy over, the organisation must demonstrate that it has real confidence in them.

Chapter Two

Planning the Year's Work

What is planning? Aims and objectives. The purposes of planning. The company plan. Six key budgets. Production budget. Cash budget. Debtors budget. Creditors budget. Capital expenditure budget. Administration budget. Checking up on the plan.

What is planning?

Planning, programming: both of these things mean looking forward, making preparations, deciding on the best course of action .

Long sight

A short-sighted businessman of any kind is bad, but a shortsighted building contractor is worse. Everything about the contractor's business needs long-sightedness and preparedness.

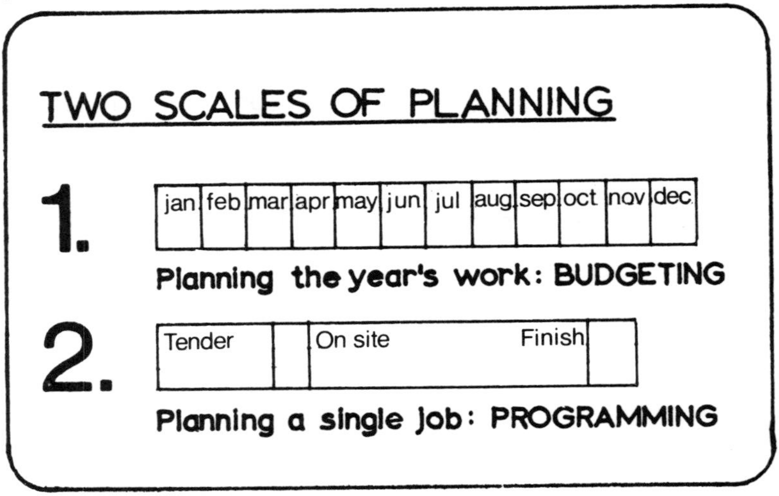

TWO SCALES OF PLANNING

1\.

jan	feb	mar	apr	may	jun	jul	aug	sep	oct	nov	dec

Planning the year's work: BUDGETING

2\.

Tender		On site	Finish	

Planning a single job: PROGRAMMING

Two scales

It is possible to divide discussion of the plan into two separate parts. One is about the large scale planning of work:

planning, over the year, for the operation of the company as a whole. The other is smaller-scale planning: planning, for a single job, the programme for one building contract.

Same foresight

In fact the same way of thinking is necessary to both types of planning: an ability to foresee what actions will be necessary and make sure that they are taken at the right time, in the right order.

Dislike of planning

The author has found that many contractors dislike the idea of planning because they associate any paper work with form filling and tax returns. They feel that time spent in the office with paper and pencil is time wasted.

We all plan

They are wrong, of course. Planning is a natural process and one which we all practise in our daily lives to some extent. We calculate roughly how much money we expect to receive each week or each month, and try to keep our weekly or monthly expenditure within these levels. We work out how much it is necessary to save if we wish to purchase a new suit and how long it will take to make.

Plan for the company

If we plan our own year's finances and activities, surely it is even more necessary to plan the year's work for our companies and organisations. Certainly, most governments think it is necessary to produce plans and budgets to achieve national growth.

How complex?

The basic problem with overall planning, as with job programming, concerns the complexity of the plan. The plan must be sufficiently carefully thought out for it to be realistic, but it should not be so complex that it takes too many hours of valuable managerial time to produce, or so detailed and complicated that nobody bothers to understand it and keep to it.

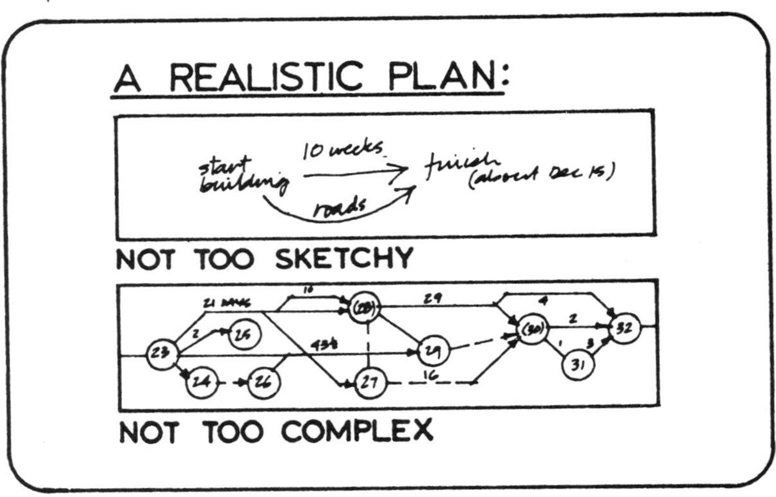

Scales of planning

A one man business can be planned 'on the back of an envelope'. At the U.K. headquarters of Costain or Taylor Woodrow (two international contractors) they use dozens of highly-paid specialists and a computer to prepare their advance plan of work. You have to suit the complexity to the scale of your business.

It gets more difficult

One thing is certain. The complexity of the problems met with by the manager of a business grow faster than the business itself grows. It is not just ten times as difficult to manage a firm with 100 employees as one with ten employees: it may be 20 or 30 times as difficult. As the firm grows to 200 or 300 employees, the problems do more than double and treble. They get to the point where one man, even if he worked 24 hours every day of the year, could not make all the decisions himself.

Planning helps

The situation would be hopeless if planning had not come to the rescue. The pace of technological and scientific change is faster today than ever before. The larger the business organisation becomes, the more difficult it will be for it to adopt new ways to meet new conditions. That is why it is very important that the small businessman should learn the modern ways of thinking about management before his business gets out of control.

Weaknesses

Of course it would be wrong to allow our enthusiasm for the idea of planning to blind us to its weaknesses. All plans are based on guesses. But guesses are not facts. You can never altogether rule out risk from business practice. However it is possible and desirable to reduce the risk to a lower level by thoughtful planning.

Minimising error

The business that never makes a mistake is like the man who never makes a mistake — it doesn't exist. Many successful businessmen have grown rich by making seven or eight right decisions out of ten. Planning, including programming as well as financial planning, is a device for improving the chances of success.

Aims and objectives

The first stage of the planning process is to decide on ***aims and objectives,*** bearing in mind your own present circumstances. This is known as a policy plan. Then, when you have decided on your aims and objectives, you need a plan over a certain fixed length of time — say a year — which, if successfully carried through, would achieve those objectives.

Write it down

One thing about a real plan is clear: ***it must be written down.*** The policy plan must not be vague: it must be expressed in real terms, in sums of money, numbers of men, amounts of material. There must be targets you can quantify and measure.

Planning targets

There will be targets for that year for all the resources of the company: not just money, but also skilled supervisors, skilled and unskilled labourers, tools and machinery.

Purposes of planning

So now we have come to the point when we can think out the purposes of planning. What are they?

1. Forward buying of capital equipment, plant and material

You need to know when to spend money — for instance, a concrete mixer must be purchased in time to cope with your expected work load.

PURPOSES OF PLANNING

1. Forward buying of capital, plant, equipment and materials.
2. Assessment of staff and labour needs, setting up a 'recruitment' policy.
3. Assessment of cash needs throughout the year.
4. Profit policy on tender prices – the profit you need to make on each job.
5. Setting TARGETS for income and expenditure.

2. Setting up a 'recruitment policy'

You need to know how many men you are likely to need and what level of ability at each moment during the year – as far as your knowledge of your forthcoming contracts will allow you to work it out. When are you likely to need to take men on, or pay them off?

3. Assessment of cash needs throughout the year

How much ready money will you need to have easily available next month, and the month after?

4. Profit policy on tender prices

This will be expressed as a percentage of turnover, say 10%. Then on a $3,000 job you would expect a profit of $300.

5. Comparing target figures

Setting target figures for income (what you will be paid for your contracting work) and expenditure (what you will pay out on men, materials, services etc.) so that you can later check your actual earnings and expenses against them. It is by comparing target figures with real figures and checking the difference ('variance') that you will get early warning that something is going wrong and, more important, *where* it is going wrong.

Where do you stand?

Before a detailed plan for the forthcoming year's work

of the firm can be worked out, it is necessary to take a long look at its present position so that you are working within a realistic framework. It is no good preparing an ambitious plan for growth if there simply aren't sufficient contracts available.

Some of the things you must have in mind when you make your company plan for the year are as follows:

1. National prospects

Are the prospects good for the industry – is there much building going on? Are the local prospects good at town or district level? Will the year see an increase in building activity or a decrease? Will it be a year of growth for you and other contractors in the country?

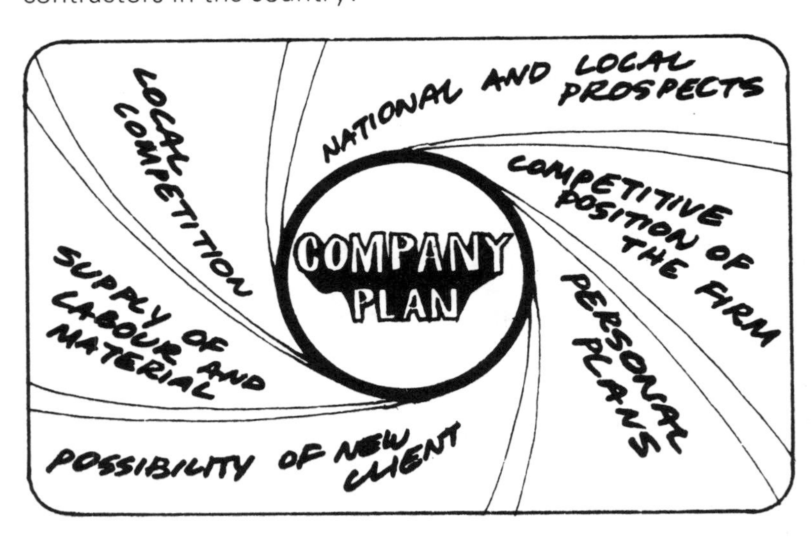

2. Competition from other contractors

How does your own firm compare with other local firms? Do you have many firms within 20 miles competing with you?

3. Competitiveness

From results of the past year's work is the firm in competitive shape? Does it need extra capital, extra plant, etc?

4. Supply position

Is labour plentiful or in short supply? What materials are becoming available, or are unobtainable? What can you expect wage levels to be like compared with last year?

5. New opportunities

What is the likelihood of finding new clients, new types of work in the coming year?

6. Your personal plans

Above all, perhaps, what do you want of the company? Or, if it is not your own personal business, but there are partners or shareholders, what do they want from the business? Do you want it to grow or stay small? Do you want immediate profit at the cost of long-term security?

Budgeting

When you have taken this kind of realistic overall view of your company's situation, you should try to work out a series of 'budgets'. The budgets will later be worked in together to form a master budget. You know how a country has a national budget announced in advance for the coming year. Perhaps you have a personal budget? For your firm you need:

Six budgets

1. Production budget
2. Cash budget
3. Debtors budget
4. Creditors budget
5. Capital expenditure budget
6. Administration budget.

1. PRODUCTION BUDGET
2. CASH BUDGET
3. DEBTOR'S BUDGET
4. CREDITOR'S BUDGET
5. CAPITAL EXPENDITURE BUDGET
6. ADMINISTRATION BUDGET

Let us think first about what people in other types of business call *sales.*

The sales target: a plan for orders

This term may sound odd in connection with a building firm. What do you sell? You sell your *services* – your organisation and building work. The sales target, then, is your plan for orders in the coming year. Write it down.

SALES TARGET

PLAN FOR ORDERS THAT YOU CAN HANDLE OVER THE NEXT YEAR

SALES PROJECTION: say $6000 is the target

month	j	f	m	a	m	j	j	a	s	o	n	d	totals
job 78.1	200	400	200										800
78.2		100	50	300	250	150							850
78.3		500	500	500									1 500
78.4				100	400	200	200	200					1 100
78.5						100	150	300	400	450	200	100	1 700
									total for year:-				5 950

Distribute it by months

Work out first what total figure of orders you might expect to obtain, and be able to handle. Be realistic. Then spread these figures between the months of the year. They will not necessarily be evenly distributed. You may expect, perhaps, to have a quiet season when the weather is unsuitable for building. Or you may expect a large school building programme to begin in the latter half of the year, on which you might expect to pick up one or two contracts.

Why you need it

You will need this rough 'guesstimate' of orders to plan your production capacity, to work out in advance when you are going to need plant, materials and labour. And your cash budget will have to be worked out to show the amount of cash in hand you will need as you go along.

Check later

If you find during the year that your actual orders are falling behind plan, you may have to make special efforts to obtain new contracts, such as reducing the percentage for profit in future tenders.

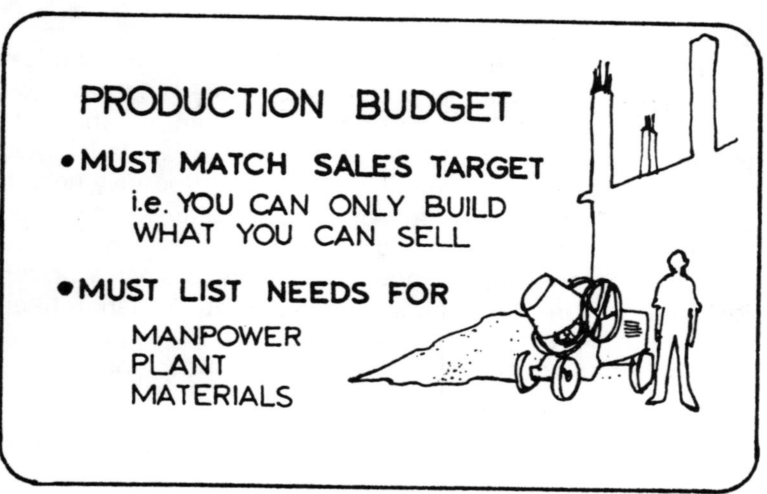

1. Production budget

Your product

What does production mean for the builder? For the manufacturer, it means what comes out of the factory; for the builder, the finished buildings.

Match order book

Of course your production budget must match your expected order book — your sales target. If you were a manufacturer, or a speculative house builder, building houses for sale, you need not match production with sales: you could be building for stock, or manufacturing goods to put in store until they could be sold at a later date.

No stock

Not so with the contractor: orders and production must be matched. It is dangerous to leave completed buildings unoccupied because they will be damaged by the weather and thieves may break in, rendering them difficult to sell later. Anyway, there are not many builders who can build without a client putting up the money in the first place.

Scheduling labour, materials and plant

The production budget means scheduling in advance your need for skilled supervisors, labour and materials and plant, and expressing them in monetary units ($), for inclusion in the cash budget.

Manpower needs

The manpower budget is part of the production budget. You will look at your expected order book and work out from one month to the next how many men, in what categories of skill, you are going to need. Then write it down. Suppose it is now the beginning of the year, you may need to make arrangements ***now***, for say June and July when you are expecting orders for some big contracts to become effective. You should look around now for a good foreman who will join you on June 1st. You should see whether there is likely to be a shortage of masons or carpenters just at the point when you are going to need to increase the number you employ.

Materials supply

Most building materials are bought only shortly in advance of taking them to the site. But some materials may be in short supply, or subject to poor delivery, and you may wish to carry stocks. Besides you may get better prices by buying in bulk. Write down roughly when you think you should buy each item, and in what quantity, if your orders plan were to be accurately fulfilled.

Plant requirements

The same, of course, applies to plant and equipment. Is it time to buy a concrete mixer, or a block-making machine now? Or would you do better to wait until mid-year, when you are expecting a lot of work, making do in the interval with hired plant? Write down what you think you should buy, either new or second-hand, and when you should obtain it.

2. Cash budget

Dangers for a building contractor

This is vital because the disasters that befall a builder more often than any other kind of business man are due to lack of ready cash. In the middle of a big job you find that, because you spent money rashly on a new lorry or on stocks of materials, or just simply because you didn't look ahead and plan the job correctly, you haven't got the money to pay the men and buy the materials to carry through what would otherwise be a profitable contract. This is the way builders go bankrupt.

CASH BUDGET

THE MONEY NEEDED TO FINANCE CONTRACTS UNTIL PAYMENT RECEIVED

	JAN	FEB	MAR	APR
JOB 1	1000	1000	1500	2000
JOB 2			1000	1000
total cash need	**1000**	**1000**	**2500**	**3000**

Idle money

Likewise, you do not want to have too much money waiting idly at the bank when it could be invested profitably in equipment, or financing additional work.

A plan for money

Try to estimate how much money you will need for each expected contract, at each stage of its completion. If you are likely to be doing more than one job at a time, add them together thus:

	January	February	March	April
Job 1 Job 2	1000 —	1000 —	1500 1000	2000 1000
Total cash needed	1000	1000	2500	3000

Foresee needs

If you are going to need a loan or an overdraft you will now be able to see this in advance, and take action. If you leave it to the last minute, you may find the bank manager unsympathetic and unable to lend you the money.

Don't drift along

The cash budget is of great importance and it is worth taking a great deal of trouble to get it right. It is the basic idea of controlling a business rather than allowing it to drift along.

Margin of error

The problem is, of course, that forecasting cannot be exact. You may be wrong in your expectation of orders for work; you may be wrong in the amount of working capital you expect to need for any one job. You should always allow a 'margin for error'. But it is better to think ahead, with only partial accuracy, than not to think at all.

3. Debtors budget

Debtors and creditors

We have seen that preparing a cash budget involves looking at the need for working capital at any moment in the coming year. Your need for working capital is, of course, influenced by the amount of credit you may have to give people with whom you have dealings and the amount of credit you take from, for instance, your materials suppliers.

Planning for debts

When planning for debts due to you, think about:

1. The actual amount of contract work you will handle in the month
2. The length of time it is likely to be before your certificates for completed work are honoured and you actually get the cash from your client(s)
3. The amount of money likely to be 'retained' by the client until some months after completion — as a guarantee of your good workmanship.

DEBTOR'S BUDGET

CONSIDER:

Note: all these 3 factors will **reduce** the cash in your bank account

1. AMOUNT OF CONTRACT WORK HANDLED IN THE MONTH
2. AMOUNT OF TIME BEFORE YOU WILL GET PAID FOR THIS WORK
3. AMOUNT OF RETENTION MONEY, AND HOW LONG WITHELD

The extent to which you are owed money in these ways is an indicator of the cash you must have in the bank to provide working capital.

4. Creditors budget

Planning for credit

When planning for what you will owe to others, think about:

1. The actual value of goods you are likely to purchase, and
2. The length of the credit period allowed by your suppliers.

CREDITOR'S BUDGET

CONSIDER:

1. ACTUAL VALUE OF GOODS TO BE PURCHASED
2. LENGTH OF CREDIT PERIOD AVAILABLE

From experience

Write it down, month by month, in advance. You can probably work it out, from your experience, as a rough percentage of your total earnings and your total expenditure at any one time.

Or you can reckon on a consistent delay:

'payment will lag behind work done by six weeks or two months as a rule.'

'I can reckon to get 30 days credit on anything I buy', and so on. But once again, leave a margin for error.

Pay up early

Don't forget, though, that you should try always to pay your debts well within the agreed credit period. A contractor should gain a reputation for prompt payment and take advantage of any discounts that are available for quick payment.

5. Capital expenditure budget

The idea of fixed assets

You must keep strict control over your expenditure on what are called 'fixed assets', that is to say, the hardware you use in your work: tools, equipment, buildings etc. You should ask yourself, before you plan to buy something, what you will ***need*** it for, what it will ***cost***, what it will ***earn*** for you, what it may ***save*** you (in labour or materials costs etc.), ***when*** you would best be advised to obtain it, and how ***long***, from the date of order, it may take before delivery.

CAPITAL EXPENDITURE BUDGET

THIS IS EXPENDITURE ON **FIXED ASSETS**

YOU MUST ASK:

WHAT YOU **NEED** IT FOR
WHAT IT WILL **COST**
WHAT IT WILL **EARN** FOR YOU
WHAT IT MAY **SAVE** FOR YOU
WHEN YOU SHOULD BUY IT
HOW LONG BEFORE IT IS DELIVERED

Check your budgets

Look at your expected work load month by month through the coming year, as shown in the Sales Budget, to see what plant and equipment is likely to be needed month by month.

Foresee your needs

Then compare it with what you know you have to hand and write down:

1. the dates on which you will need additional new items, and
2. the date you should actually order them.

Capital versus recurrent expenditure

Do not confuse expenditure on materials and wages with expenditure on capital goods like plant, vehicles and buildings. The latter last you over time, serving not just for one job but for many. They are part of your capital investment, and in money terms they are quite different from materials and wages that are allocated to one particular job.

6. Administration budget

Overheads

As the manager of your company, you are an administrator. It costs you quite a lot each year to keep the business going. You run an office and a yard. You have to maintain your plant and equipment. You have to pay, perhaps, a clerk or an estimator, you have to buy stationery, postage stamps and pay bills for lighting and other services.

ADMINISTRATION BUDGET

CALCULATE OVERHEADS:

OFFICE RUNNING COSTS
YARD RUNNING COSTS
MAINTENANCE OF PLANT & EQUIP'T

AND:

ALLOW FOR GROWTH

Foresee your overheads

It is important that you work out in advance and write down just what you think these 'overheads' are going to cost each month. If you do not plan any rapid growth in the size of your business, these may be the same each month of the year. You will need to cover the cost of these overheads by including an additional percentage for profit in each tender you prepare.

A year's plan for your company

Let's look, then, at a simple plan for a year's work. What will it consist of?

1. Production
2. Cash
3. Debtors
4. Creditors

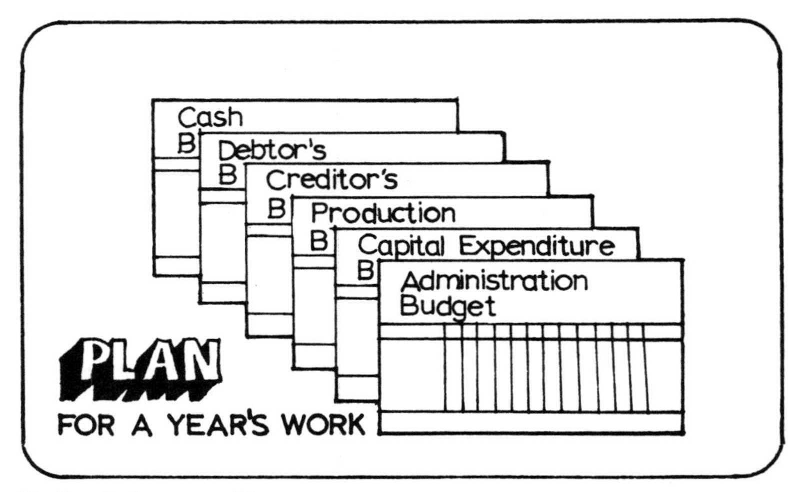

5. Capital expenditure
6. Administration.

Final task

The final task must be to cross-check all the budgets and make sure that they agree with each other and will not over-strain the resources of the company. Debtors and creditors must be kept in balance. Capital expenditure on plant and equipment must not push the cash budget beyond agreed overdraft limits. It will probably be necessary to amend some of the budgets to bring the overall plan into balance.

Now we have a plan. How do we use it?

Checking up on your plan

There is little point in spending time, effort and money on producing carefully thought out budgets in this way and then locking them away in a filing cabinet. Just as the navigator of a ship will constantly check his charts to make sure he is on course, so it is necessary for the manager to refer regularly to his budgets, programmes and plans.

Anything wrong?

If actual figures are almost exactly on target as each month goes by, then no action is needed by you. But things may go differently from the plan. For instance, orders may fall off, production may not keep up with demand or cash may get shorter than was planned. Then you should look into it and make adjustments to put matters right. Even apparently favourable factors, such as receiving more orders for work than you expected, could lead to trouble – for example, you might run short of working capital.

Keep records

You have been asked to write information down. You know that none of this planning activity is possible without accurate records. This means written records. In this way, information can be carefully filed, always up to date, and always accessible. (Keeping knowledge in your head is no good.)

KEEP WRITTEN RECORDS

1. Write everything down
2. File everything, promptly & accessibly
3. Keep Order Books
4. Keep bank statements
5. Keep petty cash and other accounting books
6. Keep receipts
7. Keep progress charts
8. Keep stock inventory

Information

Keep your bank statements, and petty cash and other accounting books. Keep receipts. Keep progress charts on your jobs. Keep an inventory of plant and materials in store.

Steer towards growth

Without this information you will be unable to run your business successfully from day to day, let alone look ahead and steer it wisely on a course of growth.

Chapter Three

Planning the Job: Preparing a Bar Chart

Planning an orderly time sequence of events on the site. Objections to programming — and the answers. Breaking down a contract into elements. Identifying key jobs. Forecasting time required. Example — bar chart for a simple bridge.

The builder's factory

For the building contractor his site is his factory. He must organise his site to produce the building in very much the same way as a manufacturer organises his factory to produce his goods. It is here on the site that he makes his profit or loss. It is here that the efficiency of his organisation is really put to the test.

Time sequence

It is on the site that time can really mean money. If work is not carefully planned, so that one job follows another in an orderly sequence, then the result will be 'waiting time', as one tradesman waits for another to complete his part of a job. While he is waiting he is not working, but he will expect to get paid. This inactivity will also delay the whole job which will result in higher wages bills, which will increase the cost of the work.

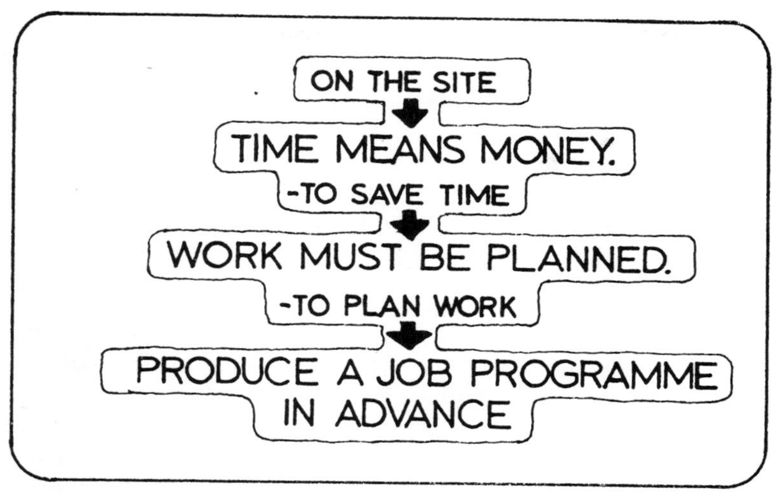

Advance planning

How is this to be avoided? The first step is to work out in advance of the start of the contract precisely how it is to be carried out. Which jobs are to be done in which order? When are materials and plant required? When should sub-contractors visit the site to carry out their jobs?

No ad hoc solution

If this is done properly, many of the snags and pitfalls that would have been met when the job was under way will be avoided. If they are met at the planning stage, they can be carefully considered and the best solution reached. But, as you know, when the foreman meets a problem on the site, he will be under pressure to come up with a quick solution, to avoid men and machinery standing idle. Thus the answer he gives may not be the best and may lead to even greater trouble at some later time.

Organising thinking

Planning and programming (or forecasting) are ways of ***organising*** your thinking about the method of carrying out contracts. As the construction industry becomes more advanced and more competitive, it becomes more important that site costs be reduced. To do this the contractor must organise and control:

- working capital
- skilled and unskilled labour
- plant, tools and equipment
- materials
- sub-contractors
- information and communications.

Considered decisions

Thus a good programme will encourage early and carefully considered decisions on all six of these vital factors in the building 'process'.

Crisis management

This is the difference between thoughtful management and 'crisis management'. Some contractors will proudly say that their minds work so quickly that they are able to deal with crises and problems as they arise. They are constantly giving instructions and orders to one man after another and are never able to give enough time to understand the problems. They think that visitors are impressed by this show of efficiency. In fact, a visitor is no more impressed by this attitude, than he would be by a taxi driver who is constantly asking the way and stopping to open the bonnet and adjust the engine.

The skill of good contract management is not to deal with crises as they arise (though this is a necessity when they do), but to try and make sure that they don't occur in the first place. This can only be done by getting early warning of problems before they develop into full scale crises.

Cannot rely on luck

There is no magic about the successful contractor. He may make a large profit on one contract by luck, but he cannot *rely* on luck. In the course of his business life he will have about the same number of chances and misfortunes as his less successful competitors. The difference is that he is *prepared* and able to act when the occasion arises.

Think ahead

To be prepared, he has to think ahead. To help him to think ahead, he should produce a job programme for all but the smallest jobs.

Planning is a technique

It can be learned with practice and constant use. It is always useful to construct a programme, as it helps to ensure that the job is thought about ahead of time. If properly used, it can tell the contractor much more than just the completion date.

Plan on paper

The important thing about a programme is that, like the year's plan for the company, it should be set down on paper. Only then can any deficiencies in the plan be clearly seen and probed.

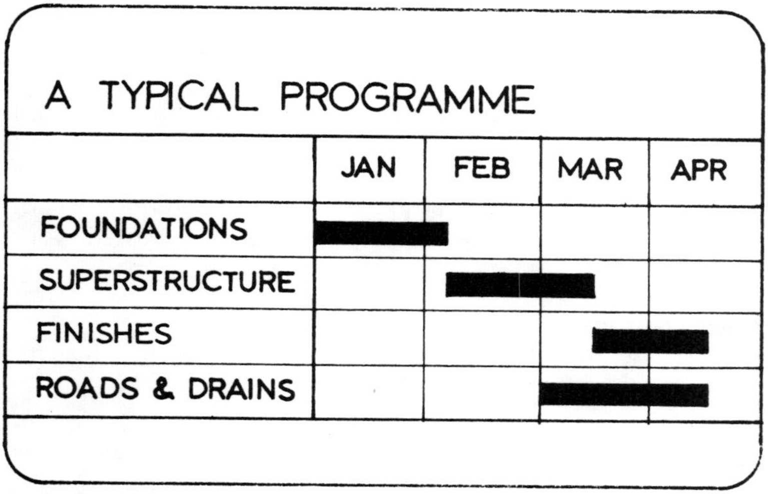

A TYPICAL PROGRAMME

	JAN	FEB	MAR	APR
FOUNDATIONS	■■■	■		
SUPERSTRUCTURE		■■	■■	
FINISHES			■	■■
ROADS & DRAINS			■■■	■■

The old way

Of course, as with budgeting, there are many old-fashioned contractors who prefer to go on in the old way with no plan.

They do not believe that it is possible to foretell the problems they may face and prefer to wait until they arise.

Objections to programming

Among the objections which they raise are:

1. A proper programme cannot be prepared when labour and material supplies are uncertain.
2. Unforseeable delays due to bad weather.
3. The time and expense of preparing the programme could be better spent.
4. The client may change his mind half way through the job, making the programme useless.

Answers

Usually, items 1 and 2 can be minimised by careful consideration of the particular area where the work is to be carried out. Also, where there are uncertainties, it is better to programme and allow for them, rather than ignore them altogether.

Item 3 is a rather foolish objection. If you compare the cost of preparing the programme with the overall value of the job, you will find that it is a tiny percentage. More important, problems will have to be faced sometime – why not face them on paper? Give yourself time to think about them carefully before the job starts rather than have to cope with them when you are in the middle of something else.

The fourth objection is a valid one but, if you point out to the client the costs of disrupting a carefully thought-out programme, he might change his mind about needless and expensive alterations. In any event, the programme might form the basis for a claim to recompense for wasted work and delay.

Positive reasons

Following the objections, here are some positive reasons for using programmes:

1. To determine the best and most economical way of carrying out the work – consistent with the contractor's resources of money, men, materials and equipment – thereby defining possible and preferred completion dates.
2. To provide a continuous work schedule or programme

for the job showing the sequence and timing of all major site operations.
3. To define the latest acceptable dates for receipt of architect's working drawings.
4. To show up some of the difficulties which might be met during the actual construction of the job.
5. To provide accurate information on what dates various materials deliveries should take place, so that they can be ordered in good time.
6. To ease the calculation of the labour force necessary to carry out the work in the time specified.
7. To indicate the periods when sub-contractors will be on site.
8. To provide information on the sequence and timing of plant and equipment requirements.
9. To show when and how much cash is required to finance the work, and as an aid to check that site costs are within estimated limits.
10. To provide an easy means of checking and controlling the progress of the job to ensure that it is up to the target stated in the tender.
11. To identify critical operations which ***must*** be done on time to meet the completion date, and show ways in which the time on site could be reduced.
12. To enable factual data to be recorded for use in estimating and planning future jobs.

Who does the planning?

The planning, in the majority of small firms, will be done by the owner himself or not at all. As the firm grows, perhaps a foreman can take over this function. It is better that this planning activity is not done by a specialist, but by the man who is responsible for the progress of the work.

Means to an end

Planning is a means to an end: the successful completion of a job. It is not an end in itself. A plan is a tool, and planning is a way of thinking about the difficulties of completing a contract — it is not a fancy wall chart.

For the small firm

As this manual is designed for the small firm or organisation, it describes only the simpler planning techniques applicable to the smaller business. As contractors grow in size, they will come to consider more advanced techniques, such as network analysis.

Checklists and bar charts

These are not really as complicated as they sound, but the needs of the majority of smaller contractors are met by the use of checklists and bar charts.

Checklists

This is more of a reminder for the manager than a planning technique. Usually checklists appear as duplicated forms with gaps to be filled in appropriate to each job.

Uses

For estimating purposes, a checklist will ensure that all the peculiarities of a site are taken into account on a site visit. Another example, for a maintenance mechanic, is to use a checklist to ensure that he has not forgotten an item of routine maintenance. Checklists can cover material deliveries, sub-contractors and starting dates for various trades.

FOREMAN'S CHECKLIST (Small job)
Job no:
Client's address
Location of site
Job description
Starting date
Duration
MATERIALS
LABOUR
PLANT
TRANSPORT
Special requirements

AN OUTLINE FOR A FOREMAN'S CHECKLIST

The bar chart

A typical bar chart gives a list of operations listed vertically on the left and a straight line or 'bar' against each showing starting and finishing dates.

JOB NO.

ACTIVITY	Week no.									
	1	2	3	4	5	6	7	8	9	10
Excavation										
Footings										
Walls										
Roof										
Finishes										

A TYPICAL BAR CHART

Simple picture

This gives a simple 'picture' of the main operations on a job and can be broken down to give much more detail if required. On a large contract (say a number of houses) there would be a master bar chart for the whole contract and individual charts for each house or group of houses.

Labour requirements

Starting with the bar chart, it is possible to calculate labour requirements throughout the construction period, dates for ordering materials and plant and requirements for working capital to finance the job.

List operations and durations

The first stage in preparing a programme or bar chart is to list all the key operations and their expected durations. Key operations are generally those that take the most time, but other key operations may be activities that involve specialist staff, hired plant or sub-contractors.

Practice

Decisions on the number of separate operations to include must be made on the basis of experience and, as with so many management techniques, practice is the road to perfection.

Breakdown by building elements

A start can be made by breaking down the overall building

process on any contract into the main building elements.

Four elements

Thus building structures may be separated into the following basic elements:

1. Foundations
2. Superstructure or shell
3. Finishings, services and equipment
4. Site works, roads, drains, pavings.

BUILDING OPERATIONS CAN BE LISTED AS:

MAIN	SUBSIDIARY
FOUNDATIONS	EXCAVATE CONCRETE BLOCKWORK TO DPC BACKFILL
SUPERSTRUCTURE	BLOCKWORK
FINISHES SERVICES & EQUIPMENT	ELECTRICAL WORK PLASTERING PLUMBING
SITE WORKS	DRAINAGE ROADS PAVING

Second step: identify key jobs

In each of these operations there will be a ***key*** operation which will take the longest time and effectively control the time taken by the rest. For instance, in the superstructure it is likely that this will be the blockwork; in the finishings it will probably be the plastering.

Third step: forecast time required

Once the time required for the key operation in each phase has been fixed (according to the size of gang employed and their output), other operations can be timed to fit, e.g. plumbing not to take longer than plastering.

Fourth step: make a chart in pencil

Finally the various items are shown on a bar chart with horizontal lines representing the time available for each particular activity, as indicated in the typical programme shown earlier in the chapter. At this stage the work should be done in *pencil*, since alterations may have to be made.

Operation lines

Where operations can take place at the same time, such as excavating foundations and external drains, they can be shown one immediately above the other. But where one must follow another, such as concreting foundations after excavating foundations, one line will start after the other finishes.

May be lucky

When this stage is finished, you may be lucky and find that the programme finishes just before the contract completion date. If it does, the programme can be accepted.

Revise if gap before completion date

But if it shows a big gap before the completion date, it gives an opportunity to revise the programme so that the labour force can be reduced.

Completion after contract date

If the programme, as often happens, shows completion ***after*** the contract completion date, then some way of reducing durations for the critical operations must be found. This may be done by increasing the labour force or introducing more plant and equipment.

Redraw

Then the programme can be redrawn (this is the reason for using pencil for the first attempt) so that the overall project time is correct.

A RESOURCE LEVELLING CHART

Job no......
RESOURCE LEVELLING
General Labourers

ACTIVITY	Week no.									
	1	2	3	4	5	6	7	8	9	10
Excavation	6	3								
Footings		3	3							
Walls			2	4	2	1				
Roof					3	2	3			
Finishes							3	4	4	2
TOTAL men/week	6	6	5	4	5	3	6	4	4	2

Check labour requirements

The final stage is to check the labour requirements set by the programme. Obviously no contractor wants to have 15 men on a site in one week, only 2 in the next, 12 in the next and 1 in the fourth week. Usually it is possible to move certain operations around so that a fairly even labour force can be maintained. This process is known as resource levelling and on a large contract can also be applied to plant (concrete mixers, dumpers, etc.).

Ink in

When the programme is drawn up satisfactorily, showing completion within the contract period with reasonably level labour and plant resources, the bars can be inked in.

Copies

A copy of the chart will be given to the site foreman and a copy will be kept in the office, with progress marked on week by week according to site reports. Any sign of work lagging behind the programme should be investigated and dealt with promptly.

Worth spending time

It is worth spending time constructing a realistic programme because it will be the manager's daily companion (and helper) until the end of the job and the final account.

Worked example

Since it has been noted that practice is the key to success in programming, the reader might like to follow through the next worked example.

Reinforced concrete bridge

The example covers the preparation of a simplified programme for a contractor who has been awarded the job of constructing a reinforced concrete road bridge over a small valley. To simplify the example, we will assume that there is no stream flowing through the valley, so the work can be carried out 'in the dry'.

Description

The bridge will run from bank A to bank D and be supported by vertical piers at B and C. After the foundations and piers have been constructed, three sets of precast units will span between the piers along AB, BC and CD. In preparing the programme, we must also allow for time to surface the road and erect safety railings.

Assumptions

It will be assumed that the precast spanning units will be

Construction of a small road bridge

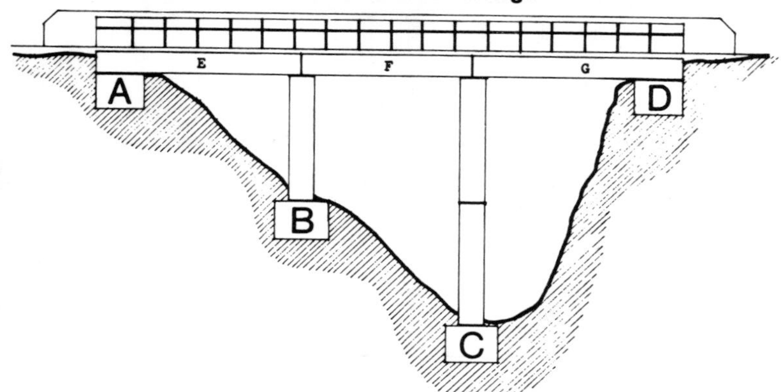

delivered to site ready for erection. Foundation pads A and D are of mass concrete but B and C contain steel reinforcement. A soil investigation has been carried out which has shown that piling will not be necessary.

Formwork

Pier B (above Pad B) can be constructed in one lift of concrete but pier C is longer and two separate lifts will be necessary. Formwork will obviously be required for the piers and a gang of carpenters will have to be employed to carry out this skilled work, but the small amount of rough formwork for foundation pads can be placed by labourers.

Three gangs

To simplify the example, numbers of employees will not be defined, and the work will be divided between the three gangs of labourers, steel fixers and carpenters.

First job: make a list of activities

As stated earlier in the chapter, the first job is to list the various activities together with their duration and the type of labour required.

Blank list of activities

To allow the reader to work through the example himself, a blank list of activities has been printed on the next page. As these are described in the text, the reader should write in the details in the 'Activity' column. An 'x' should be placed in one of the following three columns to show whether labourers, steel-fixers or carpenters carry out the task. The time taken for the task in weeks is to be shown in the last column.

Bridge Example – List of Activities

No.	*Activity*	*Labourers*	*Steel-fixers*	*Carpenters*	*No. of Work Weeks*
1					
2					
3					
4					
5					
6					
7					
8					
9					
10					
11					
12					
13					
14					
15					
16					
17					
18					
19					
20					
21					
22					
23					
24					

Key list

A completed list of activities is printed at the end of the chapter as a key so that the reader can check his working.

Activity No.1

The first activity will be ***setting-out*** for which we can allow one week. Labourers will be needed for site clearance, to act as chainmen, etc. Thus an 'x' should be placed in the 'Labourers' column and '1' entered under 'No. of Working Weeks'.

First problem

Now we are faced with our first problem. We can do nothing until we start excavating, but the major question is — where do we start? At foundation A, B, C or D?

First lesson

The solution to this problem illustrates the ***first lesson*** in practical programming. Any competent site agent or foreman will want to finish his job as quickly and cheaply as possible. He knows that the road surfacing cannot start until the slab units are laid, and the slab units cannot be laid until all the piers and foundations have been completed and allowed to cure.

Start with job taking longest time

Thus he would be wise to start with the job that will take the longest time and make that his priority. Then he will concentrate on the next longest job and programme that.

Remainder fitted in

After these jobs have been programmed, he will probably find that the remaining jobs can be fitted in.

Longest job

In this example, it is easy to see which is the longest job. Pier and foundation C involves three separate concreting operations (one pad plus two lifts of concrete for the pier), whereas B involves two separate operations (one pad and one lift for the pier). A and D can be concreted in just one operation (the foundation pad and no steel reinforcing since it is mass concrete).

Activity No.2

Thus Activity 2 is to ***excavate for foundation C.*** The time taken will depend on the shape and size of the foundation, the type of soil and the size of the labour gang. Let us say two weeks. The excavation will of course be done by the gang of labourers. Thus the reader can now fill in the details for Activity No.2.

Activity No.3

Next on the list is steel-fixing for foundation pad C. This

will vary according to the complexity of the reinforcement, but we will allow two weeks for the steel-fixing gang to do this work.

Activity No.4

The next job will be to pour concrete to this pad. Allow half a week and assume that the labourers can do this.

List remaining activities

While we are dealing with foundation C, it will be convenient to list the remaining activities covering the construction of pier C.

Activities Nos. 5, 6 and 7

Activity 5 is steel-fixing to lift one – allow two weeks. Then the gang of carpenters will be needed to fix the formwork or shuttering – allow two weeks. Finally the labourers will be able to concrete the first lift – allow half a week for this.

Activities Nos. 8, 9 and 10

That brings us to Activity No.8. The steel-fixers can then return to complete the second lift on pier C – allow two weeks. The carpenters then strip the formwork from the first lift and erect it again to hold the second lift of concrete. The final stage is Activity 10, pouring concrete to lift 2 and completing pier C.

Activities Nos. 11, 12 and 13

We have already established that foundation and pier B will prove the second longest job. Activity 11 will be the excavation of foundation, taking say two weeks. We will assume one week for steel-fixing (Activity 12) and half a week for concreting this foundation (Activity 13).

Activities Nos. 14, 15 and 16

We will allow 1½ weeks for steel-fixing pier B (Activity 14) and two weeks for the carpenters to erect shuttering (Activity 15). The pier can be completed by pouring concrete (½ week) (Activity 16).

Next operation A or D?

The remaining foundation work is to excavate and concrete foundation pads A and D. Again we have to make a choice. If we think that pier B will be completed before pier C, it would be wise to start with foundation A. The reason is that, if A is completed first, we could start the labourers laying pre-cast spanning unit E between A and B, while we are awaiting the end of the curing period for pier C. If D were completed first, the programme might be set back for a further week.

Activities Nos. 17 and 18

Thus we will make Activity 17 excavation to foundation pad A (two weeks) and Activity 18 concreting (½ week).

Activities Nos. 19 and 20

Then activities 19 and 20 will cover excavating and concreting pad D (two weeks and ½ week).

Activities Nos. 21, 22 and 23

Then we must allow for laying pre-cast spanning units E, F and G. We will assume that a mobile crane will be available and allow one week for each operation.

Activity No.24

Finally road surfacing, railings and completion will take three weeks.

Check

The reader should now have completed his list of activities, together with the type of labour required to carry them out, and the number of weeks' work involved. This can be checked against the key at the end of the chapter.

Good management

We can now proceed with producing a programme. Our objectives will be to complete the project as quickly as possible and to provide reasonably continuous employment for the various trades, so as to avoid taking men on for a week or so and then laying them off. Employees always work better when they can look forward to being employed on a site for several months, for they know that their job is worth looking after.

Bar chart: start making a programme

We start with the blank programme sheet illustrated on the next page. Across the top the time has been marked off in weeks. Since it is reasonable to estimate that the contract will take between twenty and thirty weeks, the weeks range from 0 to 30.

Bridge Example – Blank Bar Chart

	0	5	10	15	20	25	30 weeks
Pier and Foundation C							
Pier and Foundation B							
Foundation A							
Foundation D							
Precast Spanning Units, Road and Railings							
Labourers							
Steel-fixers							
Carpenters							

Mostly in weeks

Most programmes are marked off in weeks, but a small job might be divided into days, and a large civil engineering contract such as a dam might take years to complete, and be divided into monthly units for programming.

Separate into headings

Our next task is to separate out the various tasks into a number of headings. In this example, it will be convenient to use these sub-headings:

1. Pier and foundation C
2. Pier and foundation B
3. Foundation A
4. Foundation D
5. Remaining works

Foundation and pier C

We know that operations 1 to 10 can only take place one

after another, thus it will take a total of ***14½ weeks to complete pier C.*** These ten operations can be filled in one after another in the line 'Pier and Foundation C'.

Labourers, steel-fixers, carpenters

The bottom three lines named 'Labourers', 'Steel-fixers' and 'Carpenters' should be filled in to indicate which gang is occupied at any particular time during these 14½ weeks.

Other tasks

Equally these three lines will show when gangs are not employed on priority work on pier or foundation C, and could be used on other tasks.

Week 16½

No matter how quickly we complete the other pier and foundation works we will be unable to lay slabs F or G until halfway through the 16th week (allowing two weeks for curing lift 2 of pier C).

Fit in A, B and D

So if we can fit in pier and foundation work to A, B and D by using our first gangs of labourers, steel-fixers and carpenters, we will give them continuous employment and eliminate the need to take on extra labour. We also know that, if all operations up to Activity No.21 are completed by week 16½, there will be no delay to the overall project completion.

Foundation and pier B

With this in mind, we will not start excavation ***to foundation B*** until week three, when foundation C has been excavated and the gang of labourers is available.

Activity 12

Activity 12 is timed for one week but the steel-fixers will only be free for ½ week while concrete is poured to foundation C.

Avoid delays to C

We want to avoid delays to C, so they will do half the work, return to carry out Activity 5 on C, and then return to complete steel-fixing to pier B by week eight.

Labouring gang free

The labouring gang is now free, so concrete can be poured immediately (Activity 13).

Pier B complete at week 12½

We are lucky in that the remaining operations 14, 15 and 16 can follow on without delay and pier B is complete at week 12½. Allowing two weeks curing, we can start laying slabs E on week 14½, providing of course that foundation A has been finished.

Foundation A has greater priority

You will all remember that we decided that ***foundation A*** has greater priority than foundation D. It will involve the labouring gang in 2½ weeks' work. We look through their schedule and see that they are fully employed from week O to week 5½, but are then free until week 8 when they must pour concrete to foundation B.

Foundation A complete at week 8

Thus Activities 17 and 18 can be fitted in at this stage, so that foundation A will be complete at week 8.

Activity 21

Since foundation A and pier B will both be ready to take the precast spanning units at week 14½, we can schedule Activity 21 for completion by week 15½, one week before we can load pier C and carry out Activities 22 to 24.

Foundation D

The only ***foundation left is D*** which, like A, is scheduled for 2½ weeks. Our labouring gang will be able to start this work on week 8½, but will have to leave for ½ week to pour lift one of pier C (operation 7). Thus they will be able to complete pouring concrete to foundation D on week 11½.

Pre-cast spanning units, road and railings

We can now consider the ***remaining Activities*** 22 to 24. These will take five weeks in all and, since we can start at week 16½, the contract should be completed half way through week 21.

Value of programming

Thus, provided we can keep to the programme and no unforeseen snags arise, the project can be completed in 21½ weeks. This shows the value of thoughtful programming, as a careless contractor might work steadily through from A to D and find that the job would take twice as long.

The reader might like to check his work against the finished programme on the completed bar-chart, as illustrated below.

Bridge Example — Completed Bar Chart

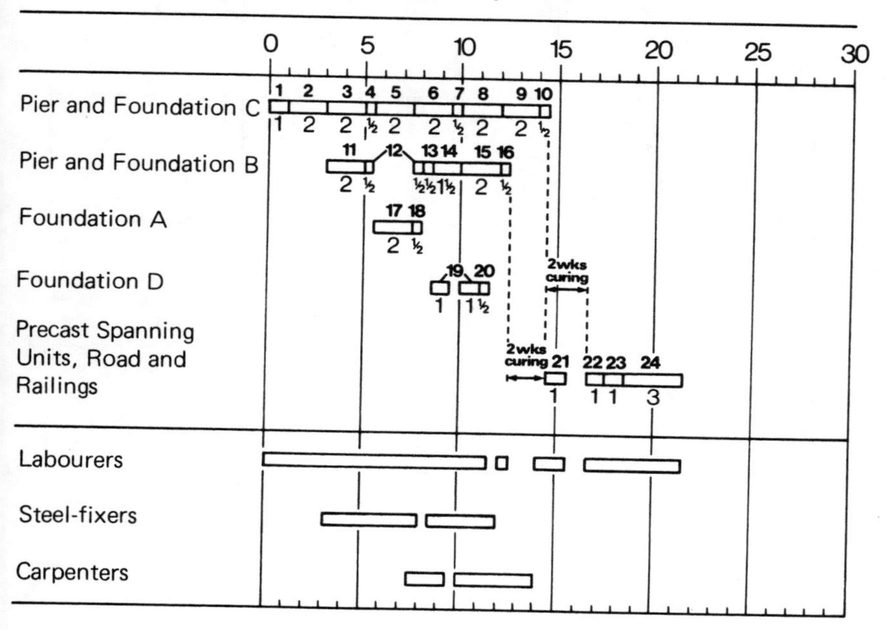

Minimum gaps

As a result of skilful programming, all three gangs are employed almost continuously, with the minimum number of gaps. The site is the builder's 'factory'. Since no sensible factory owner would think of using temporary labour, each employed for only a few days at a time to carry out a specific job, why should the building contractor?

Skilled service

Contractors should recruit good employees who fulfil their management needs with skilled and conscientious service. The foreman in this example worked well with the contractor to produce and carry through a good programme.

Bridge Example – Key to List of Activities

No.	List of Activities	Labourers	Steel-fixers	Carpenters	No. of Working Weeks
1	Setting out	x			1
2	Excavate foundation C	x			2
3	Steelfix foundation C		x		2
4	Concrete foundation C	x			½
5	Steelfix pier C – lift 1		x		2
6	Formwork pier C – lift 1			x	2
7	Concrete pier C – lift 1	x			½
8	Steelfix pier C – lift 2		x		2
9	Formwork pier C – lift 2			x	2
10	Concrete pier C – lift 2	x			½
11	Excavate foundation B	x			2
12	Steelfix foundation B		x		1
13	Concrete foundation B	x			½
14	Steelfix pier B		x		1½
15	Formwork pier B			x	2
16	Concrete pier B	x			½
17	Excavate foundation A	x			2
18	Concrete foundation A	x			½
19	Excavate foundation D	x			2
20	Concrete foundation D	x			½
21	Precast spanning units E	x			1
22	Precast spanning units F	x			1
23	Precast spanning units G	x			1
24	Road surface, railings	x			3

Chapter Four

Management Game: Job Programme and Resource Levelling

A practical exercise in the preparation of a bar chart programme and labour allocation chart for a contract covering the construction of four houses. The game also illustrates the time savings that can be achieved by the application of flow production methods to the complete contract rather than simply building one house at a time.

Learning by doing

The previous chapter provided an introduction to programming procedures related to the planning of production on the site. Following the principle of 'learning by doing', the reader might like to work through a practical example himself in the present chapter. The author hopes that this exercise will give the reader more confidence to start preparing and using programmes for jobs undertaken by his own organisation, as planning techniques are essentially practical.

Building four houses

The example at the end of the previous chapter covered (in a very simplified form) a civil engineering contract in which the contractor was required to construct a small reinforced concrete bridge. The exercise in this chapter involves a typical building contract – the construction of four houses – right through from site clearance to completion for which the contract allows 26 weeks.

Programme plus labour allocation

The reader will be given sufficient information to produce a programme and labour allocation sheet for the project, and key sheets are printed at the end of the chapter so that he can check his calculations.

One house first

Although the contract calls for the construction of four houses, the exercise will start with the preparation of a programme for building just one of these houses. The next stage will be to look at the work involved in building all four houses, and see if any time savings can be made by overlapping the operations. One house will take four months perhaps; need four houses take sixteen months? We hope not! But let us get on with the programme and see where savings of time can be made.

Exploded view

To give some idea of the way in which the contract might be split up, there is an 'exploded view' of a single house below.

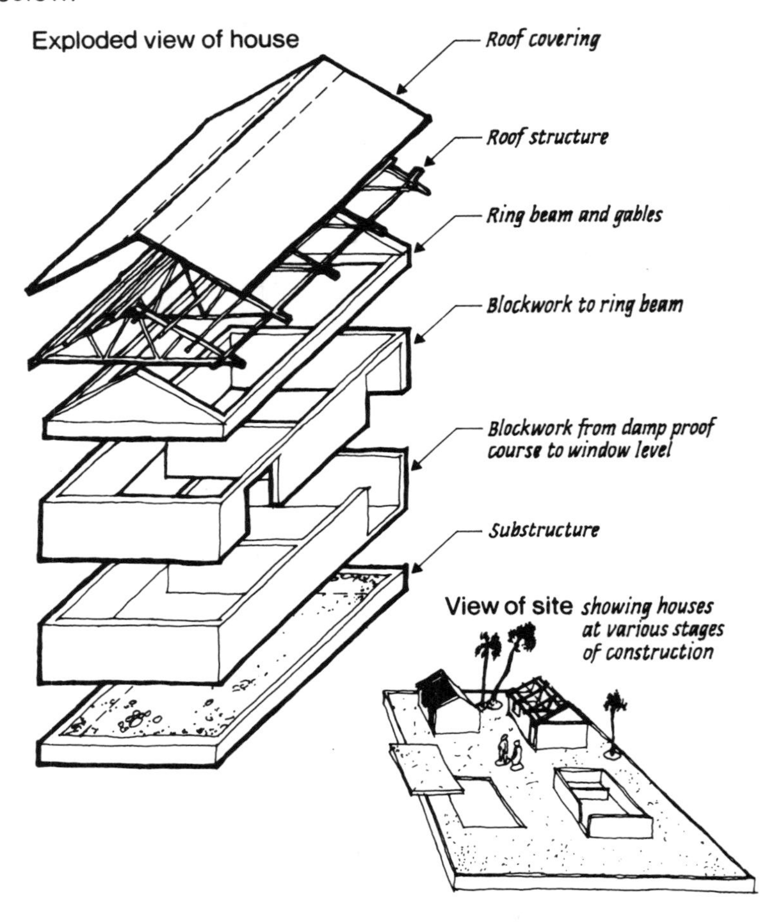

Labour schedule

The reader will recall that the first stage in programming is to list the main operations which have to be carried out.

This has been done on the labour schedule which follows below.

Labour Schedule for four Houses

Operations	Period Weeks	No of men/operation			Man-Weeks		
		Labour	Masons	Carpntr	Labours	Masons	Carpntr.
1. Site preparations							
2. layout plant & mats. Setting out							
3. Manufacture Blocks							
4. External Drainage							
5. Manufacture Roofs							
6. Excavation Conc.Founds							
7. Blockwork to DPC level							
8. Hardcore filling conc. floor							
9. Blockwork to window level							
10. Blockwork ring beam level							
11. Concrete Ring beam							
12. Roof Frame Position & Fix							
13. Roof covering							
14. Services installation							
15. Finishes Int. & External							
16 External Works (compl)							
Total man-weeks/ house							
Total man-weeks/ 4 houses							

Sixteen operations

Each of the 16 items printed on the left hand side represents one particular operation in the contract which has been sorted out from the others as being a significant activity in the small programme. Each activity is done by a group of people with some form of specialist skill.

Following pages

Details of each operation and the labour involved are given on the following pages.

First four columns

From these pages, the reader should complete the first four columns by filling in the appropriate details for each operation. Operation 14 involves sub-contractors, so, in that case, there will be no entry in Columns 2, 3 or 4.

Operation 1: SITE PREPARATION.

Operation 1: site preparation

Let us say four labourers for a week to cut down, load, and take away the trees.

Operation 2: SETTING-OUT AND LAY OUT PLANT AND MATERIALS.

Operation 2: setting-out and lay out plant and materials

The four labourers can follow on with this work for a further week.

Operation 3: MANUFACTURE BLOCKS.

Operation 3: manufacture blocks

Let us allow 2½ weeks for three labourers and one mason to include manufacture, transport and loading.

Operation 4: EXTERNAL DRAINAGE.

Operation 4: external drainage

Two labourers should cope with this in 1½ weeks.

Operation 5: MANUFACTURE ROOF.

Operation 5: manufacture roof

Remember at this stage we are just ***making*** the trusses, etc., ready to fit when the time comes – let us say two carpenters and two labourers for two weeks for each house.

Operation 6: excavation and concrete to foundations

This is basically a labouring job. Let us allow four labourers for one week to get it finished.

Operation 7: blockwork to DPC

Definitely a job for the masons. Let us allow two masons ½ a week with two labourers ***each*** to assist them.

Operation 8: HARDCORE FILLING, CONCRETE FLOOR.

Operation 8: hardcore filling, concrete floor

One mason with four labourers should get this job done in ½ a week.

Operation 9: BLOCKWORK TO WINDOW LEVEL.

Operation 9: blockwork to window level

Two masons and four labourers for one week per house.

Operation 10: blockwork to ringbeam level

The two masons and four labourers should complete this in a further week.

Operation 11: concrete ring beam

There is some shuttering as well as concreting so one mason, one carpenter and one labourer for 1½ weeks.

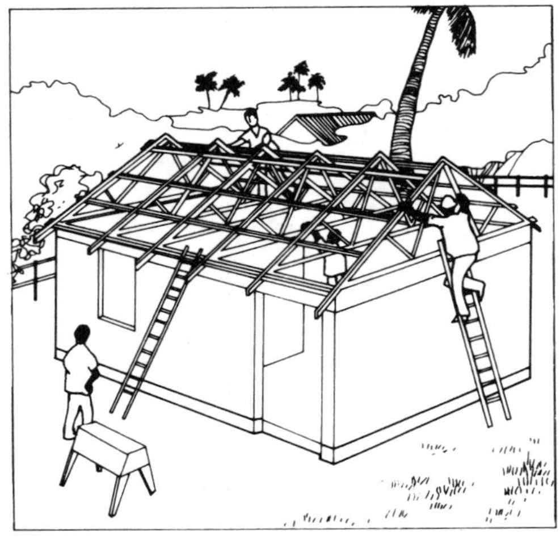

Operation 12: roof frame – position and fix

A job for the carpenters. Two carpenters and two labourers should easily do this in a week.

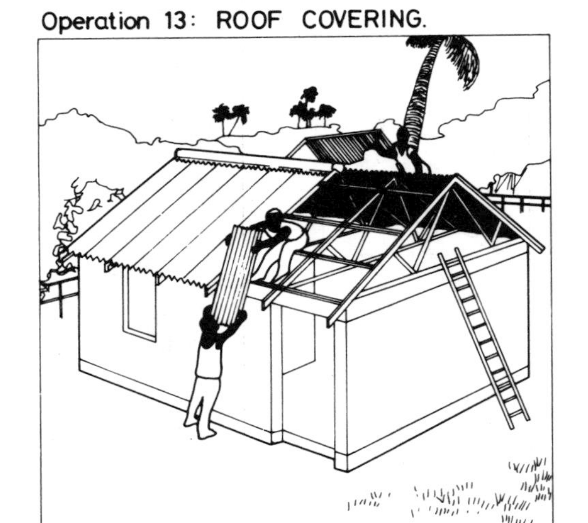

Operation 13: roof covering

One carpenter and two labourers for ½ week.

Operation 14: SERVICES INSTALLATION

Operation 14: services installation

This is a job for the sub-contractors. Let us allow them two weeks to complete.

Operation 15: FINISHES INTERNAL & EXTERNAL

Operation 15: finishes — internal and external

One carpenter and two labourers can be allowed three weeks for finishings (second fixing, painting, etc.).

Operation 16 : EXTERNAL WORKS: (COMPLETION)

Operation 16: external works (completion)

Two labourers can be allowed two weeks so that the site will be left completely neat and tidy.

Labour allocation

Now we can see from the labour schedule how many men we are going to need for each activity and for how long we estimate we are going to need them. We still have to fill in the last three columns headed 'man weeks'.

Labour content

What we really want to know for planning purposes is the ***labour content*** of each job. If it is ten man weeks we can then split it up as we wish when we come to produce the job programme. It could be five men for two weeks or, if we have plenty of time, two men could be lett to work on the job for five weeks. Either way the cost to us as contractor would be ten weeks wages. Providing we know the ***labour content*** of the activity is within the figure used in producing our estimate and we keep within that figure, we are almost certain to make a profit on that section on the contract.

Separate out the skills

You will notice that there are separate columns for labourers, masons and carpenters. This is a help to both costing and programming. We know that different wage rates are paid to different trades, that labourers cannot do tradesmen's work and that it is expensive to waste the time of a skilled tradesman on labouring.

Calculating man-weeks

We start with site preparations. This is easy. One week and four labourers makes four man-weeks in the 'labourers' column.

Setting-out

'Setting-out', etc. is the same — four man-weeks in the 'labourers' column.

Manufacture blocks

'Manufacture blocks' is a bit more complicated. First labourers — 2½ weeks for three men makes 7½ man-weeks in the 'labourers' column. Then masons — 2½ weeks for one man makes 2½ man-weeks in the 'masons' column.

Remaining activities

The reader should now complete the last three columns for the remaining activities in the same way.

Add up the men

Once the activities are complete the reader can add them up to give the total per house. At this stage leave out the first two activities, as they are carried out only once.

Total for four houses

That was for one house alone. As there will be a total of four houses to be built, these figures can be multiplied by four to give the man-weeks for the total contract, because we assume it takes four times as many man-weeks to build four houses as one. But what we will be able to do is provide continuous employment and save wasted time. This gives the total for four houses, but remember we must add the odd eight hours to the labourers column (for those first two activities).

Now let us look at the ***programme*** on the next page and work out when we are going to need to order materials and get men onto the site.

Programme sheet or bar chart

You will see the same list of activities on the left, but on the right we have a section labelled 'time' in weeks from 1 to 26 representing the contract period. Our next task is to use

Programme for Four Houses

Operations	Time in Weeks																									
	1	2	3	4	5	6	7	8	9	10	11	12	13	14	15	16	17	18	19	20	21	22	23	24	25	26
1. Site preparations																										
2. layout plant & mats. setting out																										
3. Manufacture blocks																										
4. External drainage																										
5. Manufacture roof																										
6. Excavation & Conc.Fnds.																										
7. Blockwork to DPC level																										
8. hardcore filling conc. floors																										
9. Blockwork to window level																										
10. Bwk to ring beam level																										
11. Concrete ring beam																										
12. Roof frame position & fix																										
13. Roof covering																										
14. Services installation																										
15. Finishes Int. & Ext.																										
16. External Works (comp)																										

the information contained in the labour schedule as a basis for drawing up a programme sheet or bar chart for the job.

The first house

We will start with the programme for the first house — then we will be able to see how the other three fit in afterwards.

Horizontal bar

As in the previous chapter, each operation will be represented by a horizontal bar indicating its starting and finishing date. The reader should draw in these bars according to the following instructions:

1. **Site preparation.** Obviously the first job is to clear the site (1 week).
2. **Setting out and layout plant and materials.** The second activity (1 week) to start after the first.
3. **Start manufacture of blocks.** This can start as soon as the block making machine is available, say 1½ weeks after the start (2½ weeks).

4. **External drainage.** This can also start at week 1½. You will see its duration is shown as 1½ weeks on the Labour Schedule.
5. **Manufacture roof.** This will take 2 weeks and can be started at any time so long as it is ready when the roof erection is to take place. Let us start it in week 5 and finish in week 7.
6. **Excavation and concrete foundations.** This is to take 1 week. We can start once plant and materials have been laid out in week 2.
7. **Blockwork to DPC level.** This takes ½ week. Obviously we cannot start blockwork until the trenches are concreted. So the operation takes from week 3 to 3½. Since the blocks made on site will not be cured yet, we will have to buy some in or get them from another site.
8. **Hardcore filling, concrete floor.** This is another ½ week job and can take place once operation 7 is complete.
9. **Blockwork to window level.** This activity can then follow for one week from week 4 to week 5.
10. **Blockwork to ring beam level.** Another week is required to complete blockwork to ring beam level, so blockwork on the first house will be complete by week 6, ready for the concreting of the ring beam.
11. **Concrete ring beam.** We have 1½ weeks for this operation which can follow operation 10.
12. **Roof frame – position and fix.** One week is allowed for this job which follows operation 11.
13. **Roof covering.** We can start this as soon as frame is up (½ week).
14. **Services installation.** You will remember that this job is to be carried out by sub-contractors. They cannot start until the roof is covered, so allow 2 weeks from weeks 9 to 11.
15. **Finishes – internal and external.** Three weeks is allowed for this work. There is no reason why it cannot start while the sub-contractors are in the house. Let us start it in week 10 and finish in week 13.
16. **External works (completion).** This is a 2 week operation. It can start during the last week of the previous operation, so it can start in week 12 and finish in week 14.

First house complete

We have now completed the first house. It should take

fourteen weeks, including site preparation and setting out.

Only twelve weeks left

If we only had to build one house, there would be no more problems. But of course this is only a quarter of the job and the completion date is not far away. In fact we have twenty-six weeks for the job, so there are only twelve weeks left.

Special skill

So it is impractical to build one house and then start the next, and so on. If we operated in that way, the job would take 4 x 14 = 56 weeks and heavy liquidated damages would be payable. Thus we must work out a programme which will show operations on several houses taking place at the same time. This is where the special skill of the programmer comes in, since we need to ensure that the various specialist tradesmen move from house to house in an orderly way and that there is always work available for them to do.

Last operation

What do we know? Whatever happens, the last operation on the last house must be finished by the end of the job (week 26). So let us show it from week 24 to week 26. This is the trick to this particular programming problem. We start by working through the operations for one house, then jump to the end of the job and work backwards. Of course, if the finished programme doesn't make sense or is not practical, we will have to start again and modify our method of working.

Last house

Now we have decided on the last operation on House No.4, we can see that it commences 12 weeks after the start of the last operation on House No.1. Since all four houses are exactly the same, they will take the same time to build. Thus we can go through operations 3 to 15, marking in the times for House No.4 starting 12 weeks after House No.1. Operations 1 and 2 (preparations, layout, etc.) are already dealt with for all houses in week 1 and week 2.

Houses 1 and 4 complete

We now have a programme which shows the operations for House No.1 and House No.4. But of course we haven't finished yet. We have to decide how to fit in the other two houses.

Houses 2 and 3

We know we have 12 weeks between operations on the first and last houses. It would seem to be sensible to fit the middle two evenly in between them. That would mean operations on House No.2 following 4 weeks after House No.1. Then House No.3 would follow on 4 weeks later, leaving 4 weeks until House No.4.

House No.2

Thus we can mark in all the operations (No.3 to No.16) for House No.2 starting 4 weeks after those for House No.1. At this stage the bar chart should be three-quarters complete, and it will soon be possible to analyse it more closely to see if it is realistic.

House No.3

Now we can finish our programme by marking in the operations for House No.3 starting 4 weeks after House No.2.

Programme complete

Now the programme is complete. It looks quite reasonable, but we want to be sure that the demand for labour is fairly even. We don't want to have to put 20 men on a site for one week and cut down to three or four the next. And, of course, we want all three categories of labour (labourers, masons and carpenters) to be reasonably even, not just the overall figure. We shall do this by working out a Labour Allocation for the job, so let us turn to the third sheet on the next page.

Labour allocation

This is probably the most complicated part of the exercise, so the reader should take it quite slowly. In fact, it looks something like the programme with time in weeks on a scale on the right of the sheet. But on the left we have three separate sections, one for each category of workman.

One row for each operation

If other trades were involved, there would be extra sections to cover them. If it were a simple job, where only labourers were involved, there would be just one section. Each section is broken down to give one horizontal row for each operation involved. These can be checked from the Labour Schedule and only the numbers are shown on this sheet.

Labourers' operations 1 and 2

Let us start with the first section. The first row covers

operations 1 and 2. If you look at your programme you will see that the operations run from week 0 to week 2. A check on the Labour Schedule shows four labourers are involved. Thus we show 4 and 4 in the first two boxes.

Labour Allocation (in man-weeks) for four houses

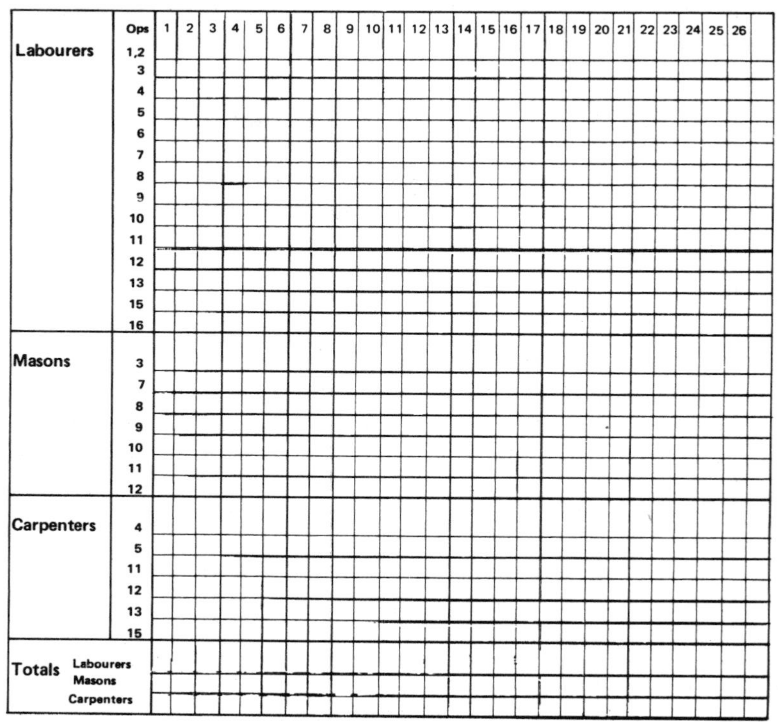

	Ops	1	2	3	4	5	6	7	8	9	10	11	12	13	14	15	16	17	18	19	20	21	22	23	24	25	26
Labourers	1,2																										
	3																										
	4																										
	5																										
	6																										
	7																										
	8																										
	9																										
	10																										
	11																										
	12																										
	13																										
	15																										
	16																										
Masons	3																										
	7																										
	8																										
	9																										
	10																										
	11																										
	12																										
Carpenters	4																										
	5																										
	11																										
	12																										
	13																										
	15																										
Totals	Labourers																										
	Masons																										
	Carpenters																										

Operation 3

Operation 3 is a bit more difficult. We see from the Labour Schedule that 3 labourer man-weeks are involved. Thus if the programme shows a full week we put the figure 3 in the appropriate box. But in week No.2 (the second box) the programme shows only half a week. Now 3 man-weeks for ½ a week gives 1½ which we show in the second box. The work goes on for two more full weeks, so we show the figure 3 in the third and fourth boxes. Then we see (from the programme) a gap of a week before House No.2 and again 1½, 3, 3 respectively. Then similarly for Houses 3 and 4.

Operation 4

The Labour Schedule shows 2 labourer man-weeks. In

week 2 only ½ a week is worked, so the figure 1 goes in the second box and 2 in the third box. Then a gap of two spaces and 1, 2 for House No.2 and so on.

Continue in same way

The reader can now continue in the same way for the remaining operations for labourers. Then similarly for masons and carpenters, so that all operations are covered (except No.14 which is done by sub-contractors).

Totals

Now we have filled in all the boxes (except for the gaps) and can work out the totals – first for the labourers. In the first column we have a total of 4. In the second we have 4, 1½ and 1 making a total of 6½. Then in the third, 3, 2 and 4 making 9. The reader can continue in the same way for the rest of the columns for labourers, then similarly for masons and carpenters. He can check his calculations against the key sheet at the end of the chapter.

A good programme

We can now see how good our programme really is. In fact, we have produced quite a practical programme with fairly even employment in all three categories. Of course, we can even out the employment pattern in practice and gradually build up to a strength of about 10 labourers, 3 masons and 3 carpenters. The labour allocation sheet also tells us when we must start to give men notice or transfer them to other jobs.

Can modify if necessary

If the Labour Allocation had ***not*** been so even, we would have had to go back to the original programme and modify it to produce a better result. This can be done ***either*** by shifting operations around where possible (e.g. manufacturing roof units earlier) ***or*** increasing or decreasing the number of men on an operation (e.g. external drainage 3 man-weeks could be 1 man for 3 weeks or 3 men for 1 week).

***Learn the skill* now**

This, then, is an example of the method of programming a job – at its most simple. You can imagine that, as the size of your contracts grow and as you handle a number of contracts at a time, the kind of programming exercise you need to do will become more complex.

It is really important for you to learn the manager's

skills of forward thinking and planning now, while your operation is small enough for you to understand the principles and practices involved.

Key to Labour Schedule

Labour Schedule for four Houses							
Operations	Period	No of men/operation			Man-Weeks		
	Weeks	Labour	Masons	Carpntr	Labours	Masons	Carpntr.
1. Site Preparations	1	4			4		
2. Layout plant & materials Setting out	1	4			4		
3. Manufacture Blocks	2½	3	1		7½	2½	
4. External Drainage	1½	2			3		
5. Manufacture Roofs	2	2		2	4		4
6. Excavation Conc.Founds	1	4			4		
7. Blockwork to DPC level	½	4	2		2	1	
8. Hardcore filling conc. floor	½	4	1		2	½	
9. Blockwork to window level	1	4	2		4	2	
10. Blockwork ring beam level	1	4	2		4	2	
11. Concrete Ring beam	1½	1	1	1	1½	1½	1½
12. Roof Frame Position & Fix	1	2		2	2		2
13. Roof covering	½	2		1	1		½
14. Services installation	2	(Sub-contractor)			(Sub-contractor)		
15. Finishes Int. & External	3	2		1	6		3
16 External Works (compl)	2	2			4		
Total man-weeks/ house					45	9½	11
Total man-weeks/ 4 houses					188	38	44

Key to Programme

Programme for four houses

Operations	1	2	3	4	5	6	7	8	9	10	11	12	13	14	15	16	17	18	19	20	21	22	23	24	25	26
1. Site preparations	=																									
2. layout plant & materials setting out		=																								
3. Manufacture blocks		=	=	=		=	=	=		=	=	=		=	=	=										
4. External drainage		=	=			=	=			=	=			=	=											
5. Manufacture roof						=	=			=	=			=	=			=	=							
6. Excavation & Conc.Fnds.			=				=				=				=											
7. Blockwork to DPC level				=				=				=				=										
8. hardcore filling conc. floors				=				=				=				=										
9. Blockwork to window level					=				=				=				=									
10. Bwk to ring beam level						=				=				=				=								
11. Concrete ring beam							=				=				=				=							
12. Roof frame position & fix								=				=				=				=						
13. Roof covering									=				=				=				=					
14. Services installation										=	=			=	=			=	=			=	=			
15. Finishes Int. & Ext.											=	=	=		=	=	=		=	=	=		=	=	=	
16. External Works (comp)													=	=			=	=			=	=			=	=

Key to Labour Allocation

Labour Allocation (in man weeks) for four houses

	Ops	1	2	3	4	5	6	7	8	9	10	11	12	13	14	15	16	17	18	19	20	21	22	23	24	25	26
Labourers	1,2	4	4																								
	3		1½	3	3		1½	3	3		1½	3	3		1½	3	3										
	4		1	2			1	2			1	2			1	2											
	5						2	2			2	2			2	2			2	2							
	6			4				4				4				4											
	7				2				2				2				2										
	8				2				2				2				2										
	9					4				4				4				4									
	10						4				4				4				4								
	11							1	½			1	½			1	½			1	½						
	12								1	1			1	1			1	1			1	1					
	13									1				1				1				1					
	15											2	2	2		2	2	2		2	2	2		2	2	2	
	16													2	2			2	2			2	2			2	2
Masons	3		½	1	1		½	1	1		½	1	1		½	1	1										
	7				1				1				1				1										
	8				½				½				½				½										
	9					2				2				2				2									
	10						2				2				2				2								
	11							1	½			1	½			1	½			1	½						
	12																										
Carpenters	4																										
	5						2	2			2	2			2	2			2	2							
	11							1	½			1	½			1	½			1	½						
	12								1	1			1	1			1	1			1	1					
	13									½				½				½				½					
	15											1	1	1		1	1	1		1	1	1		1	1	1	
Totals	Labourers	4	6½	9	7	4	8½	12	8½	6	8½	14	10½	10	10½	14	10½	10	8	5	3½	6	2	2	2	4	2
	Masons		½	1	2½	2	2½	2	3	2	2½	2	3	2	2½	2	3	2	2	1	½						
	Carpenters						2	3	1½	1½	2	4	2½	2½	2	4	2½	2½	2	4	2½	2½		1	1	1	

Chapter Five

How to Borrow Money and General Cash Flow Calculations

Borrowing and lending. Savings. Securing starting capital. Partnerships. General cash flow calculations. Collateral for loans and overdrafts. Difficulty in offering security in the contracting business.

Difficulty in finding a lender

Most businessmen need to borrow money at one time or another to finance the growth of their businesses. But for the new, small building contracting firm it is very difficult to find any kind of financial institution that is prepared to lend money. There are three reasons for this. First, because it is new. Second, because it is small. Third, because the business is building.

Newness

Banks are always wary of lending money to a new firm, unless the proprietors are already known to them as customers and have some successful business experience. This means that anyone who wishes to start a business must face up to the fact that the initial capital must come from his own savings, or funds borrowed from relatives and friends. The cure for newness, like the cure for youth, is time. After a few years successful trading, the proprietor will be able to show a reasonable track record of competent management and so be more acceptable as a borrower.

Smallness

Small firms generally find that it is more difficult to borrow money than their larger competitors. This is partly because it takes almost as much time and trouble to decide whether a small firm is credit-worthy as it does to assess the solvency of a much larger business. Even if money is made available, a higher interest rate may be charged to cover the higher proportional cost incurred by the lender.

Building as a business

The third disadvantage for the contractor comes from the

fact that he has chosen to become engaged in the building industry. Unfortunately, experience over the years in many different countries has shown that builders often head the list for bankruptcies and liquidations. This general reputation for lack of credit-worthiness means that even the better firms in the industry have to face the fact that lenders will be doubtful about their loan applications.

Competition for loans

There is never a shortage of people wanting to borrow money for a whole range of different businesses, trades and occupations. If the builder is to compete successfully with other would-be borrowers in the queue, he has to present a particularly strong case on grounds of proven ability and experience. But it will help if he knows something about how to present the case.

THE BUILDER HAS TO COMPETE WITH OTHER BORROWERS

Something special

With all this competition for loans, the building contractor has to have something special to offer if he is not to remain at the end of the queue.

Tact

It has been said that one definition of the word 'tact' is the ability to see the world from the other person's point of view. Let us therefore try to look at the building contractor from the bank manager's point of view. Like his customers, he has his own business to run (even though it may actually belong to shareholders or the government). He therefore has plenty of problems of his own, together with his own personal hopes and fears for his commercial career.

Main fear

Let us start with his main fear. Most people are aware of

the sort of businessman who turns borrowing money into a business itself. He has no intention of repaying what he has borrowed, and it is practically impossible for a lender of money to get his capital back, let alone interest on the capital sum. Most lenders have made the mistake of lending money to people of this kind at some time in their careers, and they have learned that it pays to be suspicious and cautious when faced with an application for an overdraft or a loan.

Lending money for profit

Just as the builder is in business to construct buildings for profit, the banker is in business to lend money for profit (amongst other services he provides). Money is the banker's raw material, which he must process and use to make more money if he is to succeed.

Money merchants

The banker faces the additional difficulty that the bank is itself a borrower, in the sense that it is the custodian of the money in the accounts of all its customers who are in credit. Bankers are, in a sense, money merchants who borrow money from one group of people and sell the services of that money, for a specified period and at an agreed rate of interest, to another group.

BANKS ARE,
IN EFFECT,
MONEY MERCHANTS

Integrity

The wise contractor, when deciding whether to tender for work with a private client, first attempts to assess whether he will pay his interim and final accounts promptly without any danger of bad debts. In the same way, the banker will want to know that the potential borrower is a man of integrity, who will not borrow money unless he is quite confident that he will be able to repay it at the appropriate time.

The borrower

The banker will want to study the facts and figures that are relevant to the business, but he will also want to know something about the borrower as a person. There are some men who regard it as a point of honour to stand by their word under any circumstances. Their word is their bond. But others will only keep their word if it suits them.

Knowing reputations

Good bank managers make it their business to know about the reputations of everyone with whom they are likely to have dealings in the future. They are likely to build up a network of social connections in their area, so that they will be aware of other people's judgements about local businessmen. If the local materials dealer grumbles about a new contractor who is slow to pay his monthly account, they are likely to hear about it.

Judgement

This is not to suggest that bank managers operate like spies. It just means that reputations, good or bad, cannot be easily hidden. So the best attitude for any would-be businessman is to remember that he must gradually build up a reputation for fair dealing right from the start of his career and in small transactions as well as large. The banker is wise to combine human judgement with his financial expertise and experience, when he has to make a decision on whether or not to grant an application for a loan.

Problem of finance

Having taken a look at the building contractor through the eyes of a bank manager, it is now time to look at the problem of obtaining finance for construction in a general way.

Problem of life

In a sense, the problem of finance is a part of the whole problem of life. It is a question of deciding:

1. What do you want
2. Where it is to come from.

Two sources

Money for business comes from two possible sources:

1. From savings
2. By borrowing.

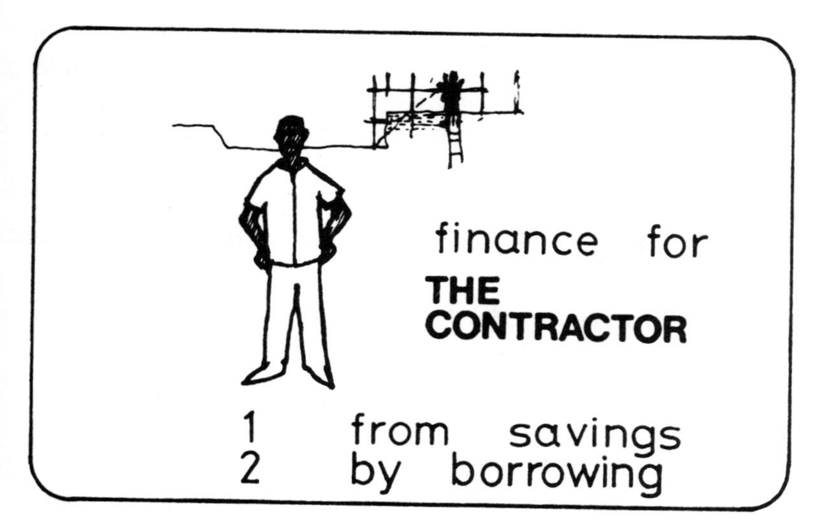

Savings

When a new business is started, it is very difficult to obtain outside finance. This is because it is at this early stage that the risk of financial loss and failure is highest. A potential lender knows that, if a new business fails, it is likely that all the money he has lent will be lost. Yet if by chance (and hopefully good management) it succeeds, he will receive only a relatively small amount of interest. Thus savings of some kind are absolutely vital if a business is to be started.

Relatives and friends

Of course it is not necessary that all the initial capital should be provided from the proprietor's own personal savings. He may have relatives or friends who are prepared to help. If they do help him with initial capital, he will have to have a clear understanding with them on the length of the loan and when it is to be repaid.

Partnership

If it is not possible to borrow money for starting capital, it may be necessary to go into partnership with someone who is prepared to put up the money. If some particular type of asset, such as office/yard premises or a truck or an excavator, is needed to start the business, it may even pay to take the owner of that asset into partnership.

Definition

The definition of a partnership is 'an association of two or more people to carry on a business as joint owners'. This means that, just as the owner of a one-man firm is responsible for that firm, including debts, all partners in a partnership have a responsibility for that firm, including debts incurred on the firm's behalf by other partners.

Advantages

The advantages of a partnership as a form of business organisation are:

1. It is relatively simple to set up
2. It is a useful method if none of the individuals have enough savings to start on their own
3. Bank managers and other potential lenders are often more willing to lend money or give credit as the risk is less.

Disadvantages

However there are also disadvantages that must be carefully considered and taken into account:

1. Partnerships are not always easy to operate
2. If one of the partners is lazy or uncooperative, the others have to carry him
3. If the partnership makes a loss resulting in a debt by the partnership, the other partners are legally responsible if one partner cannot pay.

A balance

Thus there is a balance of advantages and disadvantages, which must be weighed up by the individual businessman. The real key is trust, because unless there is trust between partners the business is bound to suffer.

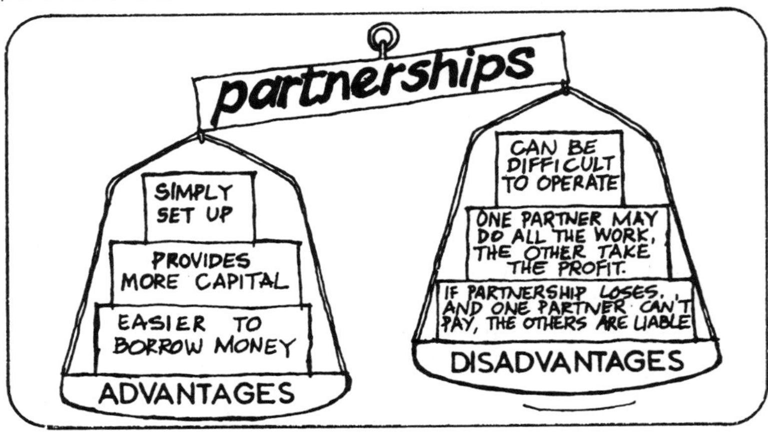

Unlimited

It must be remembered that most partnerships are unlimited. This means that a creditor of a partnership can claim against any partner individually to pay a debt. Of course, any partner paying such a debt is entitled to recover a proportion of it from the other partners. But if they do not have sufficient funds he faces a dead loss.

Useful

Providing a man can trust his partner or partners absolutely, this can be a very useful form of organisation. A partnership is quite easy to start up and can also be ended without too much difficulty if it does not work out well, so there is the further advantage of flexibility.

Not too many

One point to remember is that the number of partners should be reasonable in relation to the size of the business. If there are too many partners in a small business, they will each get so little out of it as a share in the profits that it will not be worthwhile.

Legal agreement

It is always wise to set up a partnership on the basis of a formal legal agreement, so that there can be no doubt about individual partner's rights and responsibilities. This is even true if the partners are close friends, because misunderstandings over money can strain any friendship. Items that should be included in such an agreement are illustrated below:

Legal Partnership Agreement

1. Date of agreement.
2. Names and addresses of partners.
3. Name of business, and office address.
4. Objects of business.
5. How long the contract is to last.
6. Type and amount of each partner's interest in the business.
7. Rights, powers and duties of partners.
8. Division of profits and losses.
9. Salary arrangements.
10. Conditions under which agreement can be ended.

Limited company

An alternative to setting up a partnership is the limited company, in which the proprietor and other subscribers become shareholders and their liability for debts is limited to the amount subscribed together with any guarantees that might have been given. However, setting up a limited company is fairly complicated and it is not always suitable for a small business starting with limited funds.

Other self-finance

The only form of self-finance apart from savings is from profits retained in the business after trading. This cannot apply to a new business, but it is worth bearing in mind if expansion is planned. If the proprietor of a business can limit the amount that he draws from the business in personal expenditure, the remaining share of profits will help to pay for new plant and to finance increasing levels of work in progress.

Borrowings

But a really fast-growing business can seldom be fully financed from retained profits. Few successful businessmen are fortunate enough to go right through their careers without ever having to borrow money to assist with expansion. It is far better to recognise this in advance than to be faced with a sudden crisis due to unforeseen difficulties on a major contract. Foresight is the essence of good management.

Two kinds

Borrowings can be broadly divided into two separate kinds — long term and short term. Long term borrowing usually refers to loans which will last at least for several years. Reliance on short term borrowing is more precarious because the lender can ask for his money back quite soon after the money is received by the contractor.

Use of borrowed money

The sort of borrowing that is sought depends largely on the use to which it is intended to be put. In general, fixed assets can only be funded satisfactorily with long term loans, while short term loans can cope quite satisfactorily with a need to finance current assets.

Long term borrowing

Long term borrowing is usually the most difficult to find.

Banks are not usually prepared to see their money tied up with one borrower for longer than two or three years. They have to keep a balance between borrowing and lending, and most of the money which they have available to lend is borrowed from depositors on a fairly short term basis. They have to protect their position by avoiding commitments that might tie this up for ten years or so. Thus long term borrowing must come initially from savings, although outside organisations might be more prepared to provide finance once a good trading record has been established.

Hire purchase

One way of obtaining the use of fixed assets without initial borrowing is to resort to hire purchase finance for plant and vehicles. This is a mixture of buying and hiring which can be quite useful. The contractor pays a deposit and agrees to purchase the item outright at a future date. In the meantime, he has the use of the plant, but has to go on making monthly payments until the debt is completed. At the end of the period he is able to buy the plant for a nominal sum (perhaps $1) and his obligation to the hiring company is completed.

Can be useful

The hiring company naturally includes an interest charge in its hiring rates, and it should be realised that this is rather higher than would be charged by the bank. However, hire purchase can be a very useful way of buying plant for a contractor who is sure that he can keep it profitably employed throughout the period of hire. Periods of hire vary, but most agreements are for periods of two to five years.

ANOTHER FORM OF FINANCE: HIRE PURCHASE

Short term borrowing: Circulating capital

Banks are more prepared to be of assistance to contractors

who need to finance their requirements for working capital.

Another name for working capital is circulating capital. This is because money allocated in this way does not get tied up or 'fixed' for years like money used to buy heavy plant and machinery or an office or store. It starts as cash, is used to pay wages, comes back as settlement of an interim certificate, goes to finance materials on site and returns again as cash before too long. The fact that it is likely to come back in the form of cash before very long means that the banker has more chance of seeing his loan repaid quickly.

Circulating capital for wages and materials on site

Work in progress

A builder is usually in greater need of working capital than other businessmen. This comes from the very nature of building. Unlike other forms of manufacturing, in which each individual item is produced in a matter of hours or minutes, construction projects take months or even years to complete. Although the client may make interim payments against work in progress, these payments will not usually be enough to fully finance the heavy cost of materials on site and weekly or monthly wage and salary bills.

Borrowing for specific contracts

Since working capital needs are related to the number of contracts on hand at any one time, it is often convenient to tie borrowings to the need to finance specific contracts. Bank managers sometimes prefer this, since it gives them a greater degree of confidence that the money will be repaid at the conclusion of the contract.

Calculation

A further advantage of linking borrowing to the requirements of a specific contract is that it enables the finance that is required to be calculated with a reasonable degree of accuracy. A lender of money always has more confidence in the ability of a borrower to repay if he has produced a calculation of his financial needs, since it suggests that he is working to a careful plan.

Calculating labour costs

Cash flow calculation

This 'cash flow' calculation is not as difficult as it might sound, as it is simply a way of working out how much money will be received and paid out at various stages in the contract. If this is known for each week of the contract, it is easy to see the maximum amount of money that will be needed to finance work in progress and how long this money will be needed before the client's payments put the contractor in a more favourable position.

Risk of insolvency

Cash flow calculations are important to the businessman, whether or not he is intending to borrow money. It is possible for a firm to become insolvent, even if it is sure to make good profits on its contracts, simply because it runs out of ready cash. Thus it is not enough for a contractor to produce a job programme to see how he could organise a contract. He should go on to calculate the cash flow implications to see if he can really afford to take on the additional work with his available financial resources.

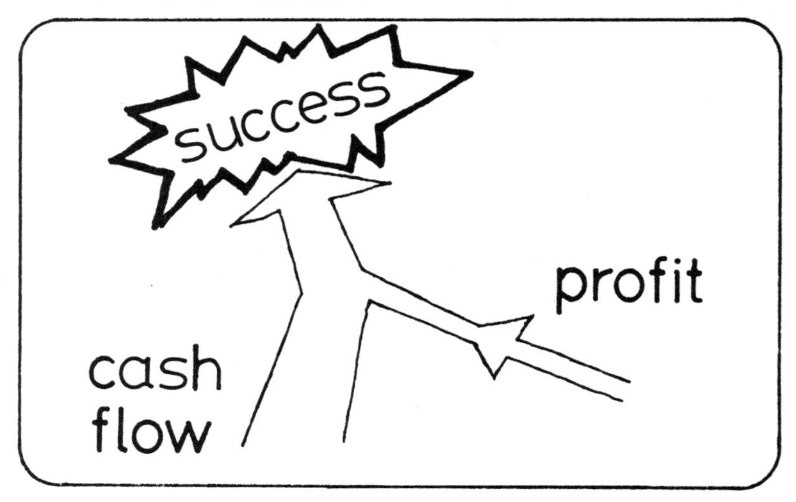

Example

To show the way to calculate cash flow on a contract, we will take the example of a contractor who has been awarded a contract which is due for completion in five months.

Start with programme

His starting point is the programme for the work, which should show when materials and plant will be needed and the manpower requirements.

Materials

Using the materials deliveries each week, it is possible to calculate weekly cash outgoings. Where credit facilities allow payment on deferred terms, the cost of the materials should be shown in the week in which payment actually takes place.

Plant and tools

Costs of plant and tools can also be calculated on a weekly basis with the cost being entered in the week in which payment is actually made. Where the contractor is able to use his own plant, such as dumpers or concrete mixers, this should bear an appropriate weekly internal hire charge to contribute to general overheads.

Wages and salaries

The labour schedule for the job should show the number of men that are likely to be employed from week to week. From a knowledge of their hourly, daily or weekly wage rate, it is a matter of simple multiplication to work out the series of weekly wage bills right through the contract.

Table

Using the above information, the next stage is to draw up a table of cash payments related to materials, plant and wages for each week of the contract. The figures for this contract might be as illustrated below:

Cash payments summary

Week No.	*Total Payments*	*Cumulative Payments*
1	250	250
2	300	550
3	600	1150
4	550	1700
5	350	2050
6	450	2500
7	450	2950
8	500	3450
9	550	4000
10	550	4550
11	650	5200
12	750	5950
13	850	6800
14	850	7650
15	750	8400
16	750	9150
17	750	9900
18	400	10300
19	150	10450
20	100	10550
21	50	10600
22	—	10600

Cumulative cash outgoings

The middle column shows the total cash outgoings for the contract during each week. The final column shows the cumulative figure, which allows the contractor to know the total amount he is likely to have to pay out in cash and cheques up to each week in the contract.

Chart

It is helpful to show these cumulative cash payments on a chart, so that the contractor has a picture of his payments over the complete period of the contract. It allows him to read off individual figures and compare one job with another.

Cumulative payments

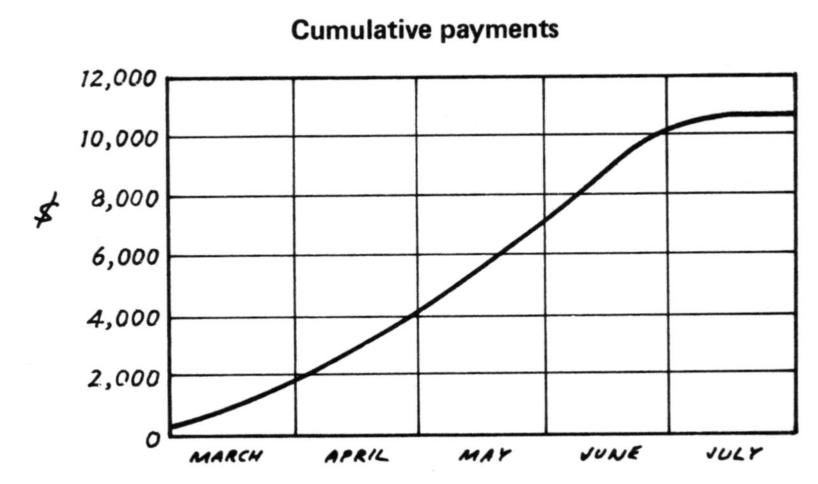

S-shape

It will be noticed that the graph itself takes the form of a flattened-out 'S'. This is the usual form of graphs of contract payments. First, payments build up slowly as the contract starts with site clearing and preparation, excavation, etc. Then the work load increases with heavy expenditure on labour and materials for the superstructure. Finally, weekly expenditure falls towards the end of the contract as labour is laid off and only painting and finishing remain for completion.

Income

Now that the expenditure has been charted, we can turn to consider income. This should be plotted on the same chart, so that the difference between cumulative expenditure and cumulative income shows the net working capital required for each week of the contract.

Interim payments

The programme can again be used as a starting point when

an income forecast is to be made. If there is a bill of quantities, measurements of work done are likely to be made monthly and payments actually received in the form of a cheque after two or three weeks. In other cases, specified proportions of the total contract sum should be paid at various stages of the work. For example, the first payment may be made on completion of site clearance and excavation. Further payments would then be made on completion of excavations, blockwork, etc.

When and how much?

Either way, the programme should give an indication of when payments are likely to be received and how much each individual payment is likely to be.

First payment

If it seems likely that the first payment will be a sum of $2000 which should be in the contractor's hands in the middle of April, this will show as a vertical line on the chart. The line will then run horizontally, making a step pattern, until the next payment.

Next payment

If the next payment to be received is expected to be about $4000 in the third week of May, a vertical line should then be drawn to bring cumulative receipts to $6000.

Sudden steps

Thus stage payments to the contractor will appear as a series of sudden steps up to the end of the contract. On most contracts, the client is entitled to hold on to 5% of the contract sum until the end of the maintenance period. Thus the final payment should take the last step up to a value of 95% of the contract sum.

Completed chart

The completed cash flow chart for the contract is illustrated below. Since the two lines represent the cumulative payments to the contractor and the payments which he has to make, the difference between the two lines gives the cash requirement to finance the contract.

Read off cash requirement

Once the chart has been drawn, it is easy to read off the cash requirement for each week of the contract. It gradually

Cash flow chart

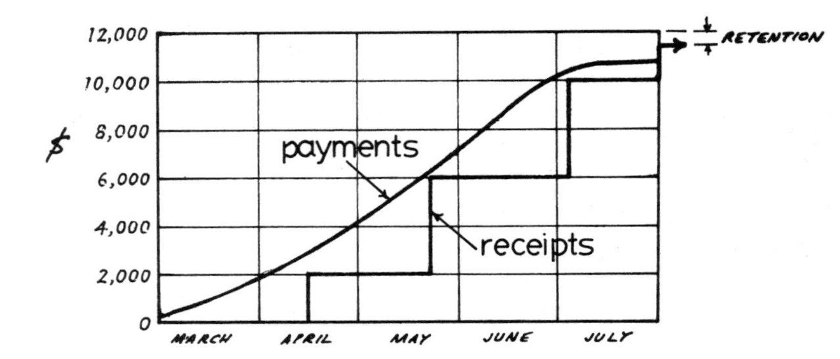

builds up to about $3000 just before the April payment comes in, but that payment brings it down below $1000. Then it builds up again to over $4000 before the May payment cuts it down to almost nothing. The maximum requirement is almost $5000 in early July.

Small positive cash balance

Half of the retention money will be released at the contract completion on 1st August, leaving a small positive cash balance. The remaining $600 retention money will not be paid over to the contractor until the end of the maintenance period.

Has $2500 – needs $2500

We will assume that the contractor has only $2500 cash available from his own resources. He therefore needs to arrange loan or overdraft facilities to cover the additional maximum requirement of $2500 to cope with the situation in early July.

Approach to bank manager

He decides to approach the bank manager to provide these facilities. The cash flow projection will help him to explain why the money is needed, and will show the bank manager that the loan should be repaid satisfactorily when the third payment is received in early July.

Ask for full amount

It is wise to ask for the full amount of $2500. It is always sensible to seek in the first place enough money to cover every reasonable contingency.

Running out of money

Too many contractors ask for less than they really need at first, and then go back to the bank when their contract is half completed and they have run out of money. They hope that the manager will be forced to lend them more at that stage so that there will be a hope of finishing the job and repaying the original loan.

Bad tactics

A borrower may even get away with these tactics once. They are bad tactics, however, and it is very unlikely that he will ever be successful in obtaining a loan in the future.

The interview

We will follow the course of the interview with the bank manager, to see the other factors that he is likely to take into consideration.

meeting the bank manager

Good first impression

The contractor will make a good first impression. The bank manager will be pleased to see the cash flow projection, because it shows that the contractor has worked out his financial needs in advance and with some care. He will also be pleased to see that the contractor is prepared to put up half of the total of $5000 from his own resources.

At least equal commitment

This covers the bank's first requirement. They are not likely to put more money into a project than the businessman himself commits. This principle gives the bank some guarantee that the businessman really believes that the project he is proposing has a genuine chance of success.

Technical competence

The next factor to be considered is the experience and technical competence of the borrower. If a man has kept a stall in the market or driven a van, the bank manager will need a great deal of convincing before he provides money for him to undertake a building contract.

Specialisation

He may even be doubtful if a firm wants to branch out into a different sector of the construction industry. For example, a firm that has previously specialised in building schools may not be equally competent if suddenly asked to construct a road.

Showing a record

The best way of demonstrating competence is to show a record of constructing smaller projects than the one currently proposed, but of the same general type.

Skilled staff

Another possibility is to show that the contractor's staff have skills and experience that are relevant, either as a result of previous employment or special training that has been provided for them.

Appropriate training

Although training starts at the top, the wise contractor sees that all his staff receive appropriate training. Both he and they will benefit from an enhanced level of skill within the organisation.

Contract documents

It is quite probable that the bank manager will want to see the contract documents, so that he can check that they include no unusual clauses.

Liquidated damages

He will be particularly interested in the provisions for 'liquidated damages': payments made by the contractor to the client to cover expenses that may be incurred if he fails to complete the job on time. This is quite reasonable, as the imposition of a fine of this kind by the client might leave the contractor with no profit margin to repay his loan.

Staff training

Bond

If the contractor has to provide a contract completion bond, he may wish this to be provided in conjunction with the bank. This would mean the bank taking on an additional potential commitment and, in this event, the manager will have to examine the contract and the contractor's previous record with special care.

Accounts

The other documents that the bank manager will wish to see are the Balance Sheet and Profit and Loss Account. He will want to check on fixed and current assets and examine the amounts of money owed to and by the business. The Profit and Loss Account (or preferably a series of them) will be needed to show that the business is sufficiently profitable to generate funds to repay the loan.

Use of profits

The bank manager will also want to see how the profits of the business have been used. If a good proportion has been left in the business, it will show that the proprietor is keen to do all he can to strengthen his firm. But if he has taken most of the profits out of the business to buy an expensive house or car, the bank manager will be less impressed. There would then be good reason to fear that the bank's money might go the same way! The bank's own records regarding the borrower's transactions might be checked for the same reason.

Security

It might be thought that any businessman who can pass all these tests deserves a loan. But bank managers are a cautious breed. They have learned to be by hard experience! They usually require some form of security as collateral for the loan.

Collateral

A collateral is a form of physical guarantee that the loan will be repaid. It is pledged to the bank for the period of the loan or overdraft facility, and the proprietor has to sign over his rights of sale to the bank. This means that, if the loan cannot be repaid, the bank can immediately sell the asset representing the collateral and repay all debts and interest outstanding from the proceeds.

Sorts of collateral

Collateral can take many forms. Stocks and shares and endowment assurance policies are very suitable, since they have a clear value and would be easy to sell for cash if necessary. Another form of acceptable security is freehold property.

Property

If the security put forward consists of property, either leasehold or freehold, the bank will put its own value on it. It will not accept the cost figure, or the contractor's estimate of the price at which he is prepared to sell. The bank's valuation will be the price it would fetch on the open market if the contractor was suddenly forced to sell.

Chapter Six

Capital Investment Decisions

Applications of mechanisation to the building site. To buy or to hire? Discounted cash flow and payback calculations. Examples illustrating calculations for evaluating capital investment decisions.

Assembly

Building is essentially a process of assembly of materials and components in a specified order and fashion to produce structures of various kinds. Most of these materials and components are large and heavy, and the output of the construction industry represents some of the most massive products of any industry.

100 to 1,000 kg/sq.m.

Even the most flimsy structures made from relatively light materials, such as bamboo, weigh at least 100 kg/sq.m. Buildings with heavy, load bearing walls of brick or concrete, but with light structural floors and roofs, like most single- and two-storey dwelling houses, weigh between 400 and 600 kg/sq.m. The more massive structure, with reinforced concrete walls, floors and roofs, can weigh up to 1,000 kg/sq.m.

Mass of materials

These figures give an indication of the mass of materials to be handled in constructing even the simplest buildings. In many countries, a house for a family of four persons weighs between 50 and 100 tons. This means that even a village of 200 families will require that 10,000 tons of materials must be brought to the site, stacked, moved around and placed.

Specialised tools and equipment

Although many building operations can be carried out quite effectively, and often more economically, by labour-intensive means, it is not surprising that some specialised tools and equipment are required for most of the assembly activities on a building site. In some cases, hand tools are sufficient, but power-driven plant and machinery is increasingly used in modern forms of construction in developing countries.

Mechanisation

Like all trends, this one can be of use to the contractor provided it is properly understood and intelligently applied. It is as foolish for a contractor to surround himself with expensive, rarely-used and inappropriate machinery as it would be to refuse even to buy a wheelbarrow on the grounds that the work could be done, as well but more expensively, with men with hand pans. It is for the contractor to take all economic, social and managerial factors into consideration and to decide on the degree of mechanisation that is appropriate for his particular enterprise.

Mixing

The first operation to be transferred, or partially transferred, to machines is mixing of concrete and mortar. Power-operated concrete mixers can assist the contractor in quality control as well as sheer output since, providing they are properly used, the concrete that is produced is of a better and more uniform standard than can usually be achieved with hand mixing.

Quality control

Quality control is always important if a contractor hopes to achieve a good reputation and so stand a good chance of a successful career in the building industry. The reason why concrete produced in concrete mixers tends to be of a better standard than that produced by hand mixing is that, even with good supervision, it is difficult to achieve the same degree of mixing time after time when the ingredients of the concrete mix are mixed by hand.

Uniformity

In mixing concrete, particularly for structural purposes, uniformity is all-important. The occasional over-strong mix can in no way compensate for the occasional over-weak mix, and poorly-mixed concrete in one vital structural element could cause the whole building to fail.

Mixers and crushers

Concrete and mortar mixers are usually quite rugged and strong. They are also reasonably simple to use, and should last for years if properly cleaned out and properly maintained. In areas where natural concreting aggregates are hard to find, a stone crusher can also be a profitable investment.

Hoists

The vertical movement of materials and components is always difficult when manpower alone is available. Thus if buildings of several storeys have to be constructed regularly, it may be worthwhile investing in some form of hoist or small crane to assist in this work.

Transport around the site

Horizontal movements are more difficult to mechanise economically, due to the fact that building sites are often uneven and there are usually many obstacles which obstruct routes through the site. Wheelbarrows are extremely useful where relatively small quantities of materials have to be shifted and the individual distances involved are reasonably short.

Site planning

Instead of thinking about ways to mechanise these horizontal movements around the site, many contractors would be better advised to think about whether they could cut out some of the movements. Every time a bag of cement or a stack of bricks or blocks is moved around the site, its effective cost to the contractor rises due to the cost of the labour involved. These costs can never be recovered by the contractor, but many of them could be quite easily eliminated by careful and thoughtful site planning in the early stages of the job.

Earth moving

In civil engineering projects there is a greater need for mechanical plant. This is particularly the case in earthmoving operations, where large quantities of material have to be excavated and moved around a site.

General transport

Trucks, particularly pick-up trucks and tippers, are of great value to a builder who operates in remote areas. They allow him to collect goods from his supplier, rather than wait for the supplier to deliver, and costs are likely to be lower as a result.

Other plant

Other plant may be required for specialist purposes. A contractor may, for instance, have to buy or hire steel sheet piles if he is awarded a contract which involves the laying of deep sewers or excavation near existing structures.

EXCAVATING NEAR EXISTING STRUCTURES

Choice

Capital investment for a contractor is thus a problem of choice, since there is seldom enough money available for him to purchase all the various types of plant that he could make use of in his various operations. Where cash is limited, he knows that more capital investment means less working capital and the risk that there will not be enough money to pay materials suppliers promptly, or even that he may be hard-pressed to find enough ready cash for wage and salary payments.

Heavy overheads

But it is wise, in any event, to think carefully before committing oneself to a capital investment decision. Plant and machinery will only pay for itself if it is in continuous use. Depreciation must always be allowed for and general overheads are quite a high proportion of total costs. These naturally go on mounting month by month even if there is no work and the plant is standing in the builder's yard. So the first question a contractor should ask himself is 'Why do I need this machine and will it be useful on future as well as on present contracts?'.

Two calculations

To some extent the answer to this question will have to be based on experience and judgement. But it is sensible for the contractor also to use calculation to supplement these qualities, and there are two factors which need to be calculated before any contractor reaches a decision on whether or not to invest:

1. Financial return
2. Impact on working capital.

Financial return

The financial return is a measure of how good the suggested investment is at making money for the contractor. It is calculated by dividing the net income produced by an asset by the cost of it. This figure is usually multiplied by 100 so that the financial return is shown in percentage terms.

Example

For example, suppose a contractor is thinking about buying a concrete mixer. He knows that it will cost him $800 and calculates that the net income it will produce will be $160. Thus the calculation is:

FINANCIAL RETURN

$$\text{Financial return} = \frac{(\text{net income})}{(\quad\text{cost}\quad)} \times 100$$

$$= \frac{(160)}{(800)} \times 100 = 20\%.$$

Attractive return

A financial return of 20% is quite attractive, and suggests that the contractor should give serious thought to making this investment. Of course the figure must be judged in relation to alternative ways of investing the money that is available, as well as his own assessment of the risk of the anticipated return not being achieved in practice.

Alternatives

If the only alternative that the contractor can see is to leave his money in a savings account at the bank earning perhaps 3%, the chance of earning 20% would be well worth looking at. But if he is offered the chance of investing in a shop which might show a return of 25 or 30% on his money, a financial return of 20% begins to look less attractive.

The risk factor

It is not just a matter of comparing figures, because it is necessary also to take into account the risk factor. Normally the rule is that the higher the risk factor, the higher the financial rate of return which we would expect on our money.

Range of rates of return

At the lower end of the range, a savings bank account with a major bank should be almost completely risk-free so a very low percentage rate of return is quite acceptable. At the higher end, a bet on an outsider in a horse race might show a profit of $200 on every $1 invested or 20,000% but with a level of risk that would only appeal to a hardened gambler.

Target rate of return

Our concrete mixer is more risky than a bank account, but (we hope) not nearly as risky as the horse. Contracting is itself a risky business, so a return on capital of 20% is not an unreasonable target.

Buying — simple but expensive

There is more than one way in which a contractor can enjoy the use of a concrete mixer on his site. The simplest way is to buy it outright, but this is also expensive and he may not have sufficient funds available. If a firm is doing well and expanding rapidly, it will need a good deal of cash for working capital so calls on fixed capital lead to an increasing and possibly dangerous squeeze on its finances.

Plant hire

Where there is a contractor's plant hire industry in operation, there is an alternative to buying plant. Plant hire can be particularly attractive when items are not needed continuously but only from time to time for particular operations. Even where there is no formal industry, it may be possible to hire specialist plant from the Ministry of Works.

Co-operation between contractors

Alternatively, a group of contractors may purchase plant on a co-operative basis or each agree to purchase one item and make it available to their colleagues on a hire basis when it is not in use.

CAN THE ANNUAL COST OF OWNED PLANT BE JUSTIFIED, OR IS HIRING MORE SUITABLE?

Buy or hire?

Where the opportunity to hire plant is available, the contractor will need to compare the costs of buying and hiring to find out which would suit him better. This decision can only be made realistically by calculating the total annual cost of owning plant (including depreciation and loan charges) and comparing this with an estimate of the hire charges that are likely to be payable over the course of a year.

Fixed and running costs

The total annual cost of owning plant is found by adding together two separate areas of cost. The first is called 'fixed cost' which includes depreciation, loan charges, insurance, taxation and any other cost item which will have to be met even if the machine never leaves the builder's yard. The second type is the 'running cost', which is related to the cost of actually using the machine when it is required. Thus:

Total annual cost = fixed costs + running costs

Running costs

Running costs usually rise steadily according to the intensity of use of the machine. In fact, some running costs, such as fuel, are related almost directly to the amount of use. If a car averages about 25 miles per gallon, then it will use 400 gallons a year if it is driven 10,000 miles during the course of a year. If the mileage the following year were to rise to 15,000, then we would expect to have to pay for 600 gallons. The same sort of relationship affects stationary plant, such as concrete mixers, although fuel usage in this case is related to the number of hours for which the plant is used.

Other running costs

Other running costs, such as repairs and maintenance, will not correspond so exactly to intensity of use. But in general, we might reasonably expect to spend twice as much on a concrete mixer which is used for 2,000 hours a year as we would on one which works for 1,000 hours a year.

Actual costs better

Where similar plant is already used in the organisation, it should be possible to base the decision on whether to buy or hire additional plant on actual costs. This is likely to provide the most realistic estimate of total annual cost. Alternatively, an experienced contractor should be able to make a reasonably accurate guess on the basis of his knowledge of local conditions.

Evaluation

Once the contractor has calculated the likely total annual cost of owning a particular machine, he is in a position to evaluate quotations for hired plant.

Annual cost of hiring

The annual cost of hiring will depend on two considerations. Firstly, the most favourable daily, weekly or monthly hiring charge must be found, if there are a number of possible hire sources. Secondly, the contractor must work out the 'utilisation', or the number of days, weeks or months for which the item is likely to be needed in an average year.

Example 1

For the first example, let us suppose a contractor is considering buying a dumper to transport materials on site. He obtains the following figures:

Annual cost of owning:	$1800
Weekly hire charge:	$75
Utilisation:	20 weeks per year.

The annual cost of hiring is found by multiplying the charge by the estimated utilisation. In this case it is:

20 x $75 = $1500

Thus in this case it would be better to hire than to buy.

Example 2

For another example, suppose the contractor needs a concrete mixer which he expects he will need for most of the year:

Annual cost of owning:	$1500
Weekly hire charge:	$44
Utilisation:	45 weeks per year

Again we multiply the weekly hire charge by the estimated utilisation, giving an annual cost of hiring:

45 x $44 = $1980

Thus in this example the advantage lies in the other direction, and it would be better to buy a concrete mixer (if funds are available) than to rely on hired plant.

Other considerations

If the figures for buying and hiring are rather close, then other considerations may be applied. If the contractor usually has several sites running at the same time, he may prefer to hire so that he could have two mixers on some occasions and none at all on others. In this way, he could cope with peak demands without any of his foremen having to wait until the firm's own plant is released from another site. Alternatively, if a mechanic is employed but his time is not fully occupied, it might be worthwhile to buy additional plant because the real cost of labour on repairs and maintenance will not be as great as it might appear in the figures. At a time of inflation, some contractors also prefer buying to hiring so that they will be able to estimate costs more accurately without having to worry about sudden increases in hire rates.

Labour-plant cost comparison

Once the most favourable cost of using plant has been calculated, it is wise to think again about whether a plant-intensive construction method is really the most suitable. It may be better to employ men than to use a machine to do the job. To find out, the contractor should compare the likely costs of labour- and plant-intensive methods.

LABOUR-PLANT COST COMPARISON...

Concrete mixer example

Taking the concrete mixer example, we assumed that it would be used for 45 weeks per year and that it would be better to buy than to hire since the estimated total annual cost of owning the mixer was $1500.

This means that the cost per working week is:

$$\frac{(1500)}{(\ 45\ )} = \$33$$

Cost of labour

Now let us suppose that the contractor estimates that four men, spending half of their time mixing concrete, could produce the same quantity and quality of mixed concrete. If the weekly wages and on-costs for a labourer work out at $15 per week, we can calculate as follows:

Cost of labour-intensive method = 4 x ½ x $15 = $30

Other factors

So in this case, the labour-intensive method might be a better and cheaper way of getting the work done than either hiring or buying. But the difference is small enough for the contractor to make his decision after taking into consideration other factors, such as convenience and quality control.

Impact on finances

If all the relative calculations and judgements suggest that it would be to his advantage to buy a large item of plant or to purchase some property, he must decide how to pay for it and what will be the impact on his finances.

Evaluating capital investment

In evaluating any form of capital investment, the contractor needs to calculate:

1. The impact on working capital
2. The likely return on the investment.

Impact on working capital

The impact of the capital investment on working capital and any possible need for loan or overdraft finance can be judged directly by the contractor in relation to his firm's balance sheet and expected cash flow pattern.

Return on investment

The return on investment is a way of calculating the financial benefit that will be enjoyed as a result of purchasing the asset, so that the contractor can judge whether it would be enough to repay him for the initial monetary outlay.

Three methods

There are three methods of calculating the return on investment:

1. Percentage return
2. Pay back

3. Discounted present value.

Percentage return method

The return is the net income or financial benefit that the asset is likely to produce in relation to its cost. Thus if a concrete block-making machine costing $1000 is likely to produce a net benefit (after allowing for depreciation and all other relevant costs) of $200, its purchase shows a percentage return of 20%.

Advantages

The main advantage of the percentage return is that it is very simple to calculate, once an accurate assessment of the likely net income has been made. It is quite a useful method of comparing several different projects, such as building a new store shed or buying a large specialist item of plant.

Disadvantages

The main disadvantage is that this method ignores the timing of the return, so it can be rather unrealistic if the financial return is delayed or varies from year to year. It also fails to tell the contractor how long the investment will take to bring in enough cash to pay for itself and start to earn real profits.

Pay back method

The 'pay back method' is designed to meet this last objection. The method uses the same forecast of future returns on the investment but in a different way, by working out how long it will take for the financial returns coming in to equal the original cost, so that the project pays back the owner for the cost of financing it. It is rather as if an uncle provides finance for his nephew to study at college on a loan basis, and would want to know how long it would take for his nephew to pay him back out of his future earnings.

Block making machine

Taking the example of the block making machine costing $1000, we remember that it will produce a benefit of $200 per year. This means that at the end of the first year it will have 'paid back' $200 to its owner. At the end of the second year it will have 'paid back' a total of $400, and so on. Thus at the end of the fifth year it will have repaid its cost of $1000 in full, so the pay back period is five years.

Another example

The pay back method can also take account of financial returns which vary from year to year. Let us suppose a businessman is thinking about opening a small brick and tile works. The initial capital investment required would be $5000. Profits would be small at first, as it would take time to persuade customers that the bricks were a good building material and to get architects to specify them on houses for their clients. Thus the anticipated net returns would be:

Year 1 — $100
Year 2 — $300
Year 3 — $600
Year 4 onwards
— $1000

Pay back in seven years

In this case we cannot take the short cut of dividing the net return into the cost figure, because it will not be stable until the fourth year. At the end of the third year only $1000 will have been repaid. But from then on he hopes that it will repay $1000 each year. This means that by the end of another four years the brick works will have repaid their owner the full $5000 that it cost him. Thus the pay back period is seven years in this case.

Advantages

The advantage of this method is that it focuses the attention of management on the importance of ensuring that capital investment yields a favourable flow of cash into the business as quickly as possible. Cash flow is particularly important in the building business, and few contractors can afford to make investments that tie up their cash for a great many years. When money is scarce, priority should be given to forms of capital investment that pay for themselves quite quickly, so that the money will soon be at the disposal of the owner again to use to finance further expansion.

Disadvantages

The first disadvantage of this method is that, like the percentage return method, it gives no indication of how long the asset is likely to continue to earn profits and also neglects any scrap value that the asset may have at the end of its life. Another serious objection is that it ignores the timing of returns from the investment.

Early returns more valuable

A further objection to both of these methods is that they value receipts equally whether they come in year 1 or in year 10. This is not very realistic since everyone would prefer to have $100 in their hands today than even the firm promise that they will receive $100 in ten year's time. Even if we had no immediate use for the $100, we know that we could invest it in a bank or savings institution so that the interest would mount up very nicely and we might be able to draw out $200 or more when ten years have passed.

Discounted cash flow

It is for this very reason that we should try to guess at the present value of the series of receipts that we are going to get at some time in the future. The technique which we can use to calculate this present value is known as 'discounted cash flow'.

Discounting formula

Mathematicians have worked out a formula which allows us to calculate the present value of $1 to be received some years in the future at a given rate of return.

If i = interest rate (%)
n = number of years

then: present value $= \left(\frac{100}{100 + i}\right)^n$

Tables

Fortunately we do not have to bother to work out this formula on every occasion, since tables are available which give the same information in a more convenient way.

Simple example

Let us start with a very simple example. Let us suppose that we have the opportunity to get a return of 10% on our investments and a dealer promises to pay us $100 a year from today. We examine the discount table and find that, for an interest rate of 10%, the one year discount factor is 0.9091.

Check formula

If we check this with the formula, we get

$$\text{Present Value} = \left(\frac{100}{100 + i}\right)^n = \left(\frac{100}{110}\right)^1 = 0.9091$$

Interest rate

Looking at this in a different way, if we had $90.91 in our hands today we would be investing it at an interest rate of 10%. Thus we would earn interest of:

$$\$90.91 \times \frac{(10)}{(100)} = 9.09$$

Thus at the end of the year we would have:

Original capital	90.91
Interest earned	9.09
Total	100.00

Tables and calculations correct

Thus our reading of the discount tables and our calculations from the formula were both correct and, at a prevailing interest rate of 10%, $1.00 in one year's time is worth $0.9091 today.

Previous examples

We will now examine the previous examples from the point of view of the discounted present value. Again an interest rate of 10% will be assumed, so we can extract discount factors from the tables as follows:

Year 1 – 0.909	Year 6 – 0.564
Year 2 – 0.826	Year 7 – 0.513
Year 3 – 0.751	Year 8 – 0.466
Year 4 – 0.682	Year 9 – 0.424
Year 5 – 0.621	Year 10 – 0.386

Block making machine example

We will first look again at the proposed purchase of a block making machine, assuming that it will have a useful life of ten years. We had calculated that it would produce a return of $200 per year on the basis of a cost of $1000. We obtain the present value of the income for each year by multiplying it by the appropriate discount factor. The calculation is best done in a tabular form as follows:

Present Value Calculation — Block making machine			
Assumed interest rate: 10%			
Year	*Income*	*Discount Factor*	*Present Value*
1	200	0.909	181
2	200	0.826	165
3	200	0.751	150
4	200	0.682	136
5	200	0.621	124
6	200	0.564	112
7	200	0.513	102
8	200	0.466	93
9	200	0.424	85
10	200	0.386	77
		Total	$1225

Favourable investment

This means that the block making machine appears to be a potentially favourable investment, since it will bring in income with a present value of $1225 for a cost of $1000.

Brickworks

As a second example, we will return to the proposed investment by a local businessman in a small brick and tile works. The data were as follows:

Initial investment:		$5000
Income:	Year 1:	$ 100
	Year 2:	$ 300
	Year 3:	$ 600
	Year 4 onwards:	$1000

Interest rate 10%

Again we will assume an interest rate of 10%, so the calculation is:

Present Value Calculation — Brickworks
Assumed interest rate: 10%

Year	*Income*	*Discount Factor*	*Present Value*
1	100	0.909	91
2	300	0.826	248
3	600	0.751	450
4	1000	0.682	682
5	1000	0.621	621
6	1000	0.564	564
7	1000	0.513	513
8	1000	0.466	466
9	1000	0.424	424
10	1000	0.386	386
		Total	$4445

Looks less favourable

Thus this investment looks less favourable at first sight, with a present value of $4445 against a cost of $5000. But it is fair to say that the brick and tile works will go on producing income beyond the initial ten year period which we have examined, and this income will increase the present value figure usefully.

Changing interest rate

The calculations are naturally affected considerably by the level of interest chosen or assumed. A higher interest rate favours short term investments which bring in money quickly so that it can be re-invested as soon as possible, even if the flow of cash does not last very long. A lower interest rate favours investments like the brickworks which take some time to reach their full profit-earning capacity.

Brickworks at 5% interest rate

To show the effect of a lower interest rate we will re-calculate the brickworks example on a 5% interest rate:

Present Value Calculation – Brickworks

Assumed interest rate: 5%

Year	*Income*	*Discount Factor*	*Present Value*
1	100	0.952	95
2	300	0.907	272
3	600	0.863	532
4	1000	0.822	822
5	1000	0.783	783
6	1000	0.746	746
7	1000	0.711	711
8	1000	0.677	677
9	1000	0.645	645
10	1000	0.614	614
		Total	$5897

Now more favourable

As we expected, we now find that the potential investment in the brickworks looks more favourable than before, with a present value of $5897 against a cost of $5000.

Advantages and disadvantages

Although discount calculations are more difficult and tedious than the other two methods, they do have the advantage of taking into account the cost of money. This applies even when the businessman has money already available for investment and so does not need to borrow. The discounted cash flow method allows him to evaluate alternative potential investments which he might be considering in a more realistic way. However, it remains for him to use his experience to do the all-important job of evaluating the risks that might be involved in each of these alternatives.

Chapter Seven

Billing Procedures; Control of Debtors

Clients and customers. Applying for interim and final certificates. Retention money. Extras and additional work. Dealing with dayworks claims. Control of debtors.

The most important people

If you were to ask any group of businessmen "who are the most important group of people to your businesses?", you would get a whole variety of replies. Some would say "myself and my partners, because we make all the key decisions". Others might answer "my foremen, because they make all the day to day decisions on the site". Others might say "the craftsmen — carpenters, plumbers and masons — because of their specialist skills". Others might even refer to the manufacturers and suppliers of materials, since no work could go on if there were no materials and components to build with.

Customers and clients

They would all be wrong! The most important group of people to any business are its customers and clients. Without them there would be no business at all. So they really are the VIPs.

Architects and engineers

In the building business, many clients, particularly private clients, employ their own professional advisers. They represent the client's interest in site discussions and advise the client on contract procedure and payments. More important still, they advise their clients on whether to include an individual contractor on a tender list or not. Thus the architects and engineers who advise and act on behalf of their clients must be regarded by the building contractor as yet another group of VIPs.

Public sector also

This attitude should not only be shown by the private

enterprise contractor who builds for profit. It should also apply to the direct labour organisations in the public sector. They must remember that they are not building for their own pleasure or satisfaction, but to produce assets which will be used and enjoyed by the whole community or sectors of it. Thus they have to have their customers and clients. The school builder should remember the interests of teachers and children. The builder of a health centre should remember that it will probably be used by his own family, his friends and neighbours. They are the people who will be the judges of his competence and craftsmanship.

Treat as VIPs

Since clients and their advisers are VIPs, they must be treated as such. If none of them invite him to tender for another contract, they will not be hurt because the contractor will have many competitors, but his business will be finished. Sometimes clients will be unreasonable (in the contractor's eyes) when they ask for faulty work to be replaced. Sometimes they will be selfish and greedy (in the contractor's eyes) when they refuse to certify extra work or dayworks. In these circumstances, the wise and mature contractor tries to explain and persuade, but never loses his temper.

Contractor's staff

The contractor's staff should be instructed to show the same intelligent approach. Clients can get rudeness and lack of consideration anywhere — to start with they probably get it from their own customers. They are entitled to expect that once a contract has been signed, their contractor will get on with the job straight away, and complete it on time to the required standards of quality and finish. Neither the contractor nor his staff are doing the client a favour. The contractor is adding to his turnover and (hopefully) his profit. His staff are doing the jobs they are being paid to do. Neither would receive this remuneration if the client had not shown sufficient confidence to give them responsibility for the contract.

The risk business

It is very rare for any contract to be completely trouble-free. It has already been stated that the contractor is very much in the risk business. His risks start when he forecasts future costs and forms the estimate which will be the basis for his tender. They continue when the contract has been signed and work starts on the site. They finish only when the work has been handed over to the client, accepted as satisfactory and the final account and retention money have been paid.

Must expect trouble

Thus the contractor must be realistic and expect trouble — although he should of course do his very best to foresee it and avoid it. If the problem is not his responsibility and entitles him to a contract variation or payment for extra works, this must be drawn to the attention of the client's representative.

Do not share troubles

But where the trouble is his own responsibility, he cannot expect sympathy. If he forgets to order cement long enough in advance and the work is held up, he is to blame. If his foreman resigns to join a competitor, it is his problem. Nothing will be gained by sharing this sort of problem with the client. He probably has enough of his own problems anyway. If there is a crisis, it is up to the contractor and his staff to work twice as hard to keep the work going and not lose time. Only if there is no possible way of avoiding delay or inconvenience should the client be told, and then the contractor should not expect sympathy — reluctant acceptance of the position by the client is all he can hope for.

Don't always give in

This does not mean that a contractor should always give in to his clients. If he is sure of his facts, he should demand his rights under the contract if extra work is requested or required — although in a respectful way. Firmness should always be combined with tact.

A variety of clients

A contracting business will be stronger if it works for a variety of clients. Although large contracts are always tempting, they often show a smaller profit margin and can make the contractor more vulnerable. Not only can contractual and climatic risks be more serious but, if the client were to go bankrupt or even to simply take a personal dislike to the contractor and refuse to allow him to tender for future work, the firm could suddenly be without work. Thus ten $10,000 contracts for separate clients can represent a healthier workload than one $100,000 prestige project for a single customer

New clients

Since clients and customers are so important, it is well worth while for the contractor to give some thought to ways in which he can attract new clients as well as holding on to those which he already has. Compared to other businessmen, the building contractor has a problem in getting new clients and customers. He cannot put his wares on a market stall or open a shop. Even newspaper advertisements will not help to get any work other than maintenance and petty works. Advertisements in technical magazines can sometimes help. But most contractors only get invited to tender for new work on the basis of personal recommendations. Thus the best advertising agent any contractor can have is a satisfied client.

THE BEST ADVERTISING AGENT:
A SATISFIED CLIENT

The client's point of view

So try to look at your work from the client's point of view. What does he expect from his contractor? Here are some of the questions that might be asked. It is for the individual contractor to provide the answers!

Do your clients end with a building which merely just about satisfies the conditions of contract or is it something to be pleased with and proud of?

What do they say about you as a builder?

Do they comment favourably on your standards of quality and finish?

Are they completely happy with the service you provide?

Have you a reputation for starting jobs promptly and finishing on time?

Do they believe that you possess the most important asset for any businessman — a reputation for integrity?

Clients talk

Always remember that your clients (past and present)

talk. In fact they often talk to people who are your potential clients in the future. Thus it is vital to build up a good name with anyone and everyone who can give you business in the future.

Sites and plant

Another way of giving a good impression to potential clients is to keep sites tidy and to keep plant and vehicles clean and well-maintained. If a lorry is rusty, dirty and falling apart, the builder would be well-adivsed to hide his responsibility, by putting his name in the smallest letters possible. Too often we see such lorries proudly displaying the name of the builder, and the public gains the reasonable impression that he has low standards and will be quite happy to produce shoddy and careless work.

CLEAN AND WELL-MAINTAINED PLANT AND VEHICLES ARE AN EXCELLENT ADVERT-ISEMENT FOR THE FIRM.

Cheap advertising

We seldom see a successful firm with shoddy plant. Successful builders know that the cost of a few tins of paint and a few hours of labourer's time in washing down plant once a week is cheap advertising as well as good maintenance procedure.

Client must keep his side of the bargain

Of course the contractor's problems are not at an end when he has found a client. In a sense that is when his problems really start. He has to use his managerial and technical skills to carry out the work to his client's satisfaction and fulfil his side of the contract. But he also has to use his financial skill and his commercial judgement to make sure that the client keeps his side of the bargain that the contract represents.

When will they pay?

What the contractor most values in a client is a reputation

for prompt payment for work once it is certified by the client's representative. Work carried out but not paid for is a dead loss as far as the contractor is concerned. Even if he achieves a profit of 10% on turnover, one unpaid account for $1,000 can only be recovered by tendering for and carrying out as much as $10,000 of good profitable work. Thus any wise contractor always asks himself one question before agreeing to carry out any work for a new and unknown private client. That question is "when will they pay?".

A reference

In these circumstances, it is just as reasonable for the contractor to require a reference from a bank manager or established business firm to confirm the client's financial ability to pay accounts when they become due as for the client to require a reference to confirm the contractor's competence. The best way to deal with bad debts is to avoid them in the first place, because there are so many examples of promising small firms being crippled or destroyed by failing to receive payment for work carried out in good faith.

Accounting and control

So the first stage in dealing with clients is to attract them to allow you to tender for work. The second stage is to ensure that the client has the money to pay for progress and final payments as the work is done. The third stage is to set up accounting and control procedures so that clients and customers can be billed for work promptly and accurately when payments are due, and a check can be made on any debts that are overdue.

Battle for survival

This third stage is also important in the battle for survival as a contractor. It is a stage which is foolishly neglected by too many small contracting businesses.

Satisfactory completion

They know that it is important that their tender should be accepted. They also know that the work has to be carried out efficiently on the site either to an agreed interim payment stage or to completion before they can expect to receive any money. But they then put off the task of preparing an account to send to the architect or his client. What they fail to realise is that it is not until that account has been accepted and honoured, with the client's cheque duly paid into the bank, that the contract can be said to be satisfactorily completed from the contractor's point of view.

A proper account

The contractor should remember that the client will not pay until a proper account has been received, just as the contractor would not pay his materials supplier until he had received a properly detailed and laid out statement and invoices. If it is scribbled on a dirty sheet of paper, with wrong unit prices, incorrect arithmetic and includes hidden additional charges, there will be no great rush to do anything about it.

Time and trouble

In fact, the client would be quite justified in sending it back and refusing even to consider it until it was submitted in an acceptable form. It will mean more difficulties for the contractor in trying to sort it out, and may well lead to arguments and unpleasantness resulting in the loss of an opportunity to tender for future profitable work. Thus the initial preparation of an account is an activity that justifies time and trouble on the part of the contractor.

PREPARING THE ACCOUNT: IT MEANS TIME AND TROUBLE

The final account

The final account is particularly important from this point of view. It is the last communication that the architect and client will receive from the contractor. If it is well-presented, with quantities, prices and extras all clearly stated, it will leave a good impression. Besides this, it will make prompt payment more likely, thereby releasing cash to finance other contracts.

Cycle of cash to cash

Most construction companies rely heavily on internally

generated cash flow to provide funds for expansion. The object on the site should be to turn materials rapidly into work in progress and into completed work. That completed work should then be turned into debtors as quickly as possible and the debtors turned into cash. This process of turning debtors into cash is the last link in the chain, which makes up the complete cycle of cash to cash. If a company makes a profit of 10% on turnover, completion of the cycle of cash to cash in four months instead of six months would mean that working capital can earn 30% a year instead of 20%.

Progress payments

On all but the smallest building or civil engineering jobs, the contract makes some form of provision for partial payment of the total contract amount to the contractor at specified periods as the work progresses. This saves the contractor from having to borrow very large sums to cover work in progress, and is to the advantage of the client since the loan interest on such sums would be reflected in much higher tenders.

Monthly intervals usually

The usual provision is for payments at monthly intervals, although this can vary with the type of contract. In some types of 'lump sum' contracts, payment is made at various stages of completion. For example 10% of the contract sum may be paid when the foundations of a building are complete, a further 25% when the walls are finished, a further 25% on completion of the roof, and the remainder when finishes and internal services have been dealt with.

Certificate or account

It is usually the responsibility of the contractor to prepare the account or certificate, showing the amount that he feels is due to him in recompense for the work that has been done to date. This includes the cost of work actually completed valued at agreed unit or bulk rates, together with materials on the site which are to be incorporated in the work. The account is really a request for payment to the client, which will be examined carefully by the architect or engineer who represents him to ensure that it is realistic.

Valuation

The method of valuation of work done depends on the

nature of the contract. If the contract is based on a priced bill of quantities, the interim account will have to be based on the quantity of work completed on the date when the account is prepared. For example if the tender included an item (say Item 24) for 100 cu.m. of concrete and the contractor had tendered $20 per cu.m., the tender item would be shown at $2,000. If at the interim stage 25 cu.m. had been mixed and placed, the interim account would include:

Item		*Quantity*	*Rate*	$
24	Concrete, 1:2:4	25 cu.m.	$20	500

Lump sum contracts

In lump sum contracts with provision for monthly measurement, the valuation would be based on estimated percentages of the major components of the whole contract. For example, if a contract consisted of the construction of roads, sewers and a building, the valuation would be based on percentage completion as below:

Building	*Contract Amount* $	*Estimated Completion* %	*Interim Valuation* $
Roads	12,500	10	1,250
Sewers	5,000	60	3,000
Building	17,500	20	3,500
	35,000	Total	7,750

Materials on site

Materials on site are valued at cost, and supplier's invoices should be available to confirm unit prices. Components and prefabricated items are similarly valued at the amounts billed by the appropriate subcontractors. The contractor should remember that once the materials have been delivered to the site they belong to the client, so they can only be removed with the express permission of the client or his representative.

Retention money

Most contracts permit the client to withhold a certain proportion of each interim payment as a guarantee that the contractor will stay on the job until the work is properly finished. This proportion is usually 10% of the accumulated value of the work certified in the interim account, although other proportions are sometimes specified.

Maintenance period

When the work is accepted as complete by the client's representative, there is usually a requirement that the contractor should put right any faults that may appear during some agreed succeeding period. This 'maintenance period' is often six months, but may be twelve months or up to the end of the rainy season in tropical areas. Thus, although half of the retention money is usually released at agreed completion, the remaining 5% is retained by the client until all faults are corrected at the end of the maintenance period. If the contractor leaves the job or fails to put right the faults in a reasonable time, he is liable to forfeit the retention money which may be used by the client to pay another contractor to do the necessary work.

Extras and additional work

In all but the simplest contracts some work has to be carried out in addition to that allowed for and priced by the contractor in the contract document. For example, the foundations may have to be deeper than expected due to difficult ground conditions or one of the rooms in a house extended due to a change of mind by the owner. In some cases, it may be possible to price the additional work on the basis of rates included in the bill of quantities. For example, there may be a priced rate per cubic metre for excavation or per square metre for block work. If these rates are available and the contract variation is not very large, it is just a question of re-measuring the work on completion and multiplying by the appropriate unit rate.

Shrewd tendering

This is where a shrewd contractor can gain an advantage in preparing his tender. If he guesses that the actual quantity for some particular item, such as excavation in rock, will be much larger than shown in the bill of quantities, he knows it will pay to boost his profit margin on that unit price. Thus his tender will stay competitive, but he will make a bigger profit when the item is re-measured at the conclusion of the contract. Equally, if a contractor guesses that the actual quantity will be less than that shown in the bill of quantities, the profit margin can be cut since the tender will be made much more competitive but little profit will be lost. However, this procedure can be very risky since a contractor who guesses wrongly can end up losing money if a variation is ordered.

Methodical approach pays

The contractor will find that it pays to adopt a methodical approach to the recording and calculation of extra work. It is not the responsibility of the client or his representative to ensure that the contractor receives payment for work carried out in addition to that allowed for in the contract. They merely have to certify what is correctly and fairly claimed. If he has neglected to record the work at the time and forgets what is involved by the time the contract is completed, then there is no way for the costs incurred to be recovered.

Extra works orders

The basis of the methodical approach is good documentation. If the contractor has been wise enough to ask the client's representative to sign a numbered extra works order to certify additional work before it is commenced, there will be little danger of any extra work being forgotten. The numbering of the orders will help to identify any gaps in the sequence. Extra works orders also save time in calculating the cost of variations by showing how it is to be measured and priced.

Proper documentation

The additional minutes spent in proper documentation in the early stages while the extra work is being done will save hours of research (or guesswork) at the end of the job. If there is any doubt, it is sensible to discuss quantities and prices with the architect or quantity surveyor before the account is prepared and sent in. It also helps to get extra works payments agreed soon after they are done, so that there will be less to argue about at the time of the final account. The contractor should always aim to make his interim and final accounts as clear, simple and non-contentious as possible, so that the client's representative will have no reason to waste time over them and delays in issuing a certificate will be minimised.

Daywork jobs

Where there are no comparable rates in the bill of quantities and the work cannot be defined sufficiently in advance, extra works are usually paid on the basis of 'dayworks'. In this method of payment, the actual amount of staff time, plant time and materials is recorded day by day, and the cost is reimbursed to the contractor together with a specified percentage addition to cover overheads and profit. Since the percentage addition is not large, the contractor can easily lose money on daywork jobs if he is not very careful to see that no relevant charges are forgotten.

All documents available

The important thing to remember is that all documents must be available to justify the daywork charge, and these should be endorsed by the client's representative and dated as soon as they are available. The documents needed for each daywork account are:

a. Time sheets (signed by clerk of works or inspector), together with rates of pay;
b. Purchase orders for materials, together with delivery notes and invoices;
c. Orders and supplier's invoices for hired plant;
d. List of contractor's own plant and tools used on the job and hourly or daily rates (which should be comparable to hire charges).

Write up dayworks records regularly

Any carelessness in filing or keeping these vouchers could lead to under-recovery of costs. Thus daywork records should be written up regularly and if the job is expected to continue for several weeks, the daywork accounts should be built up (and agreed) weekly.

Sales journal

After the interim or final account has been submitted, the final operation is to enter the amounts owing by clients in a Sales Journal, together with the date on which the account was sent. The journal will thus provide a useful check on debtors and payments overdue, so that reminders can be sent promptly. Of course, the contractor should make a point of entering cheques received promptly in the journal, as the client will be justifiably annoyed to receive a demand for payment which has already been made.

Control of debtors

When a client or customer owes money to a contractor he becomes a debtor. Providing the client pays his debts on time there is no problem. But this will not always be the case, and the contractor needs to think out his policy on debt collection and the management of accounts receivable.

Four key factors

This policy will be based in the case of any particular client or potential debt problem on the contractor's assessment of four key factors:

a. Risk of eventual default;
b. Impact on budgeted cash flow;
c. Cost of working capital tied up;
d. Profit potential related to cost and risk.

Example

For example, a trusted client who is almost certain to pay up eventually and allows the contractor a good profit margin on his work may be allowed more time to pay than a new customer who is getting a cheap job and is unlikely to require any more work in the near future. But if there is a cash flow problem or high interest rates are being incurred on a loan or overdraft, the contractor may be forced to take a much tougher line on all outstanding debts.

Regular check

The contractor would be wise to keep a check on the size of the debtors figure regularly, as it can rise steadily and eat up more and more essential working capital.

Days' work equivalent

One simple check is to calculate the number of days' work that the cash tied up in debtors represents. This is calculated by dividing the debtors figure by the average value of work carried out in a working day (calculated as an average figure for the previous month, six months or year).

Example

For example suppose:

Debtors = $12,000
Average daily value = $ 500

$$\text{Days' Work Equivalent} = \frac{(12{,}000)}{(\ 500\ )} = 24 \text{ days}$$

Further investigation

When the 'Day's Work Equivalent' figure remains stable or steadily reduces, it is likely that the debtors item is under control. There is cause for further investigation, however, if it starts to rise for some reason. What would be particularly worrying would be for some debts to remain unpaid for a very long time, since it might be a sign that the money they represented might never be recovered.

Length of time outstanding

To avoid these long term debts building up unnoticed, it is also useful to keep a regular check on the length of time that each debt has been outstanding, grouped as follows:

a. Up to 30 days;
b. 1-2 months;
c. 2-3 months;
d. Over 3 months.

Example

It is helpful for analysis, if a percentage column is added as in the following example:

Debts outstanding

	Amount $	%
Up to 30 days	6,600	55
1-2 months	2,640	22
2-3 months	960	8
Over 3 months	1,800	15
	$12,000	100

Examine the trend

Again it is helpful to examine the trend compared with previous analysis as well as the actual percentages themselves. In this case the debts totalling $1,800 outstanding over three months should be closely examined to see if there is any danger of default.

Chapter Eight

Calculating Cash Flow on a Contract

Building with money. Pre-commencement costs. Preliminaries. Interim accounts. Payment procedures. Cash flow forecasts, Calculating payments and receipts. Worked example of contract cash flow projection. Combining contract projections. Cash forecasts for jobbing work — example. Ways to improve cash flow. Cost comparison of loans and overdrafts.

Building with money

We know that a contractor uses materials and plant to build, and he needs to arrange in advance for sufficient of the right materials and plant to be available on the site when they are needed. His job programme should help with this. But he also needs money to build, and the quantity of working capital needed to finance work in progress will also vary from time to time as the contract progresses.

Programme helps

Here again a properly thought out job programme can help the contractor to schedule his resources efficiently and effectively. His programme will help him to work out when materials should be delivered to the site and, by adding on a an appropriate credit period, he can calculate when he will have to send a cheque to settle with his suppliers. The programme should also give an indication of the amount of work that is likely to be completed as each interim certificate is issued and, by adding an allowance for the estimated time taken by the client to honour the certificate and make payment, the contractor can calculate probable receipts as the contract progresses.

Pre-commencement costs

The contractor starts to incur expense before his men and his plant even arrive at the site. He may be required to provide a bond as guarantee that he will carry out the work, which would mean either putting up a sum of money in cash or paying a premium to a bank or insurance company which would agree to bear the risk in the event of a failure by the contractor. In addition, the contractor may be required to take out various types of insurance policy to protect himself (and possibly the client) against claims by employees or third parties.

Preliminaries

Further costs are involved before the contractor can get on with productive (and profit-earning) work. He will probably have to erect a site office for himself and another for the Clerk of Works. A shed or store will be necessary to protect materials and tools from the weather and also from theft. Fences and barricades to protect the public may be necessary if excavations are to take place in roads or footpaths, and it is usually best to fence in a building site to discourage possible trespassers. Further costs will be incurred in installing temporary water, electrical, sanitary and telephone services. All these costs will have to be financed by the contractor initially, as there will be no opportunity to recover them from the client until he is in a position to claim his first interim certificate.

Commencement

Even the commencement of actual work on the project will further worsen the cash flow position at first. Wages will have to be paid weekly, and some materials will have to be paid for in cash if no credit arrangements have been negotiated. Even when credit arrangements are available, the supplier may offer an attractive discount for cash settlement and the wise contractor would wish to be in a position to take advantage of this.

First interim account

Thus costs continue to build up in the first few weeks after the work has commenced. Then, usually about a month after commencement, the time comes for the first interim account to be prepared and submitted to the client. Even then, the money will not become immediately available. The client or his representative will have to check the account, and any errors will have to be corrected. Since it is very much in the interest of the contractor to reduce the time for checking to a minimum, he will be wise to ensure that every account which he submits is as accurate and clear as possible. When the account has been checked, a further two to three weeks may be needed for the client to complete his own payment procedures and forward his cheque to the contractor.

Payment procedures

Payment procedures vary from client to client. Even some government departments can be slow to make money available to their contractors, and the reputations of private clients for prompt payment or otherwise can vary enormously. The experienced contractor will know which clients can be relied on for quick payment of accounts, and this will be a factor which will lead him to submit more competitive tenders for their work than for potential customers who have a less reliable reputation.

Cash deficit rises again

While the contractor is waiting anxiously for the first payment from his client, his outgoings continue to mount up. Thus the first payment is not likely to put the contract 'into the black' and, although it will cut the cash deficit on the contract when it arrives, that deficit will start to rise again as credit accounts are settled and further wage and cash payments have to be made.

Expenses exceed income

The cash deficit will be relieved again when the second interim payment is received, but a contractor's expenses on a project often continue to exceed income for the biggest part of the job. Where profit margins are particularly low, he may not even begin to show a cash surplus until the retention money is released.

Working capital

This cash deficit is something a contractor has to allow for in planning his business. It means that he must either have sufficient working capital to finance what he will need to cover the deficit from day to day or arrange loan or overdraft facilities so that cash will be available when it is needed.

Cash flow

The day to day movement of money into and out of the business is known as the cash flow. It is not the same thing as profit, but is just as important as profit. We know that if a business does not operate at a profit its cash resources will be steadily reduced and the stage will be reached when it will not be able to continue unless it is subsidised. But not every businessman realises that he can be awarded profitable contracts and still go bankrupt. If a business suddenly runs out of cash, it is too late to negotiate emergency loans and the employees and creditors will want to be paid immediately — not when the contracts are complete and the client has paid the contractor for work done.

Cash flow forecasts

With these dangers in mind, the contractor needs to know his working capital requirements well in advance. This must start with a forecast of the likely cash flow on each major contract. It would be better for a contractor to miss the opportunity to carry out a contract than to take it on and find — too late — that his firm would run out of money as a result.

The method

The method of producing a cash flow forecast for a particular contract is to estimate separately the most likely dates on which payments will have to be made in respect of costs incurred and the schedule of progress payments which will be received from the client. Once the times and amounts of payments and receipts have been forecast, the difference between cumulative payments and receipts gives the cash flow position at any particular time.

A simple example

It is convenient to tabulate the figures for ease of calculation, and a simple example of the form the calculation might take is shown below. It is always best to put figures neatly in the form of a table when calculations have to be made, as much time can be lost due to mistakes when the figures are scribbled on a piece of paper. This applies particularly when additions have to be made horizontally as well as vertically.

Cash required rises steadily

The example is quite typical for a building contract in that the cash required to finance the contract is shown to build up gradually as the contract proceeds to a peak (in this case) of $1500. This represents about a quarter of the contract sum, and this proportion is itself not unusual. An experienced contractor will have a good idea of the propor-

	Cash Flow Forecast — Example				
	Payments	Receipts	Cum. Payments	Cum. Reciepts	Cash Position
January	200	—	200	—	–200
February	750	—	950	—	–950
March	1050	600	2000	600	–1400
April	1100	1000	3100	1600	–1500
May	900	1400	4000	3000	–1000
June	850	1000	4850	4000	–850
July	400	1200	5250	5200	–50
August	100	850	5350	6050	+700

tion that usually applies on his contracts, and this will give him a quick guide to the work he will be able to take on with a given amount of working capital.

Then improves

After reaching the peak of $1500 the cash position steadily improves until it has almost reached a break-even position in July. The excess of receipts over payments in August means that it then starts to bring cash into the firm.

Monthly intervals

Although cash flow forecasts can be produced for any time interval, even daily if required, for most purposes a monthly forecast is suitable. Intervals of more than a month might mean that a critical peak in cash requirements would not be foreseen while weekly or daily forecasts would involve very detailed and complicated calculations.

Calculating payments and receipts

The previous example started at the stage when monthly payments and receipts had been calculated. In the next case, this earlier calculation is carried out for the contract to build four houses dealt with in Chapter Four. It is this calculation which is usually the most difficult since the contractor has to estimate when payments for each stage of the work will be made by him to employees, sub-contractors and suppliers, and when appropriate stage payments will be received from the client.

Judgement and experience

As with preparing a programme for the project, these estimates must be made on the basis of judgement and experience. But, despite the uncertainties, most competent contractors can produce reasonably accurate cash flow forecasts by working through the following procedure in a thoughtful and methodical manner.

Four houses example

For this example, we will return to the contract to build four houses which was introduced in Chapter Four. In that chapter a programme was produced which would lead to completion of the work in 26 weeks (or six months). The programme is as follows:

Programme for four houses

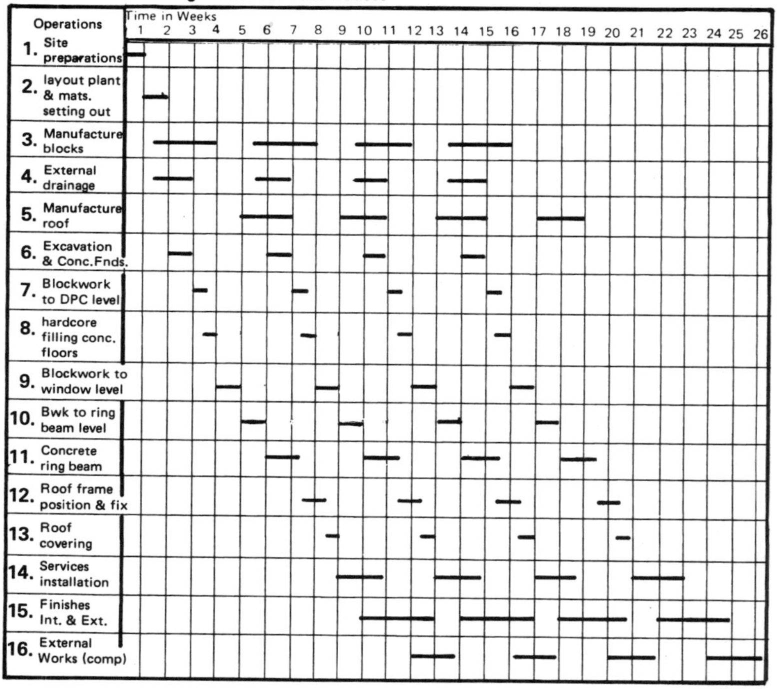

Operations

Time in Weeks

1 2 3 4 5 6 7 8 9 10 11 12 13 14 15 16 17 18 19 20 21 22 23 24 25 26

1. Site preparations
2. layout plant & mats. setting out
3. Manufacture blocks
4. External drainage
5. Manufacture roof
6. Excavation & Conc.Fnds.
7. Blockwork to DPC level
8. hardcore filling conc. floors
9. Blockwork to window level
10. Bwk to ring beam level
11. Concrete ring beam
12. Roof frame position & fix
13. Roof covering
14. Services installation
15. Finishes Int. & Ext.
16. External Works (comp)

Monthly interval

Although the programme has been prepared on the basis of a weekly interval, we need to base our cash flow calculations on a monthly interval.

Cost figures

The first stage in working from the programme to the cash flow forecast is to estimate cost figures for all of the operations listed in the programme. We need to know not just the amount of money that will have to be paid out to cover costs on each activity, but also ***when*** it will have to be paid. In some cases the whole payment will be made in a single month but, where credit may be made available or the activity will spread over several weeks, the costs will be split up over a number of months. Although perfect accuracy will not be achieved, an experienced builder should be able to estimate costs and payment dates with reasonable accuracy.

Four Houses – Estimated Payment Dates

Operation	*Cost*	*Payable in equal parts in months*
1. Site preparations	1000	Jan, Feb
2. Layout plant and materials	500	Jan
3. Manufacture blocks	1200	Feb, March, April, May
4. External drainage	900	Jan, Feb, March, April
5. Manufacture roofs	1000	Feb, March, April, May
6. Foundations	800	Feb, March, April, May
7. Blockwork to D.P.C.	600	Jan, Feb, March, April
8. Hardcore filling, concrete floor	800	Jan, Feb, March, April
9. Blockwork to window level	600	Feb, March, April, May
10. Blockwork to ring-beam level	400	Feb, March, April, May
11. Concrete ring-beam	1200	Feb, March, April, May
12. Position and fix roof frame	900	March, April, May
13. Roof covering	1500	March, April, May
14. Services installation	3000	March, April, May, June
15. Finishes	1200	March, April, May, June
16. External works	400	March, April, May, June
	16,000	

Assumptions

It will be assumed that the six month contract will run from January to June, and cost figures are assumed to have been abstracted from the estimate for each of the operations 1-16.

These cost estimates are listed above, and in the third column the contractor has made a separate forecast of when the cash will have to be expended. It is important to remember that, for this purpose, the months given are those when the cash is expected to be actually paid out – which may be later than the month in which the cost is incurred if materials or services are made available on credit.

Total payments

The next stage is to separate the payments due in each month for each operation as in the following table. The totals for each month show the amount of money that the contractor is likely to have to pay out in respect of the contract in that month.

Operations	*Total*	*Jan*	*Feb*	*Mar*	*Apr*	*May*	*Jun*
1. Site preparations	1000	500	500				
2. Setting out layout plant & mats.	500	500					
3. Manufacture blocks	1200		300	300	300	300	
4. External drainage	900	225	225	225	225		
5. Manufacture roofs	1000		250	250	250	250	
6. Excavation, concrete founds.	800		200	200	200	200	
7. Blockwork to DPC level	600	150	150	150	150		
8. Hardcore filling, concrete floor	800	200	200	200	200		
9. Blockwork to window level	600		150	150	150	150	
10. Blockwork ring-beam level	400		100	100	100	100	
11. Concrete ring-beam	1200		300	300	300	300	
12. Roof frame position & fix	900			300	300	300	
13. Roof covering	1500			500	500	500	
14. Services installation	3000			750	750	750	750
15. Finishes Int. & External	1200			300	300	300	300
16. External works (completion)	400			100	100	100	100
	16000	1575	2375	3825	3825	3250	1150

Now forecast receipts

Now that the total payments likely to be made in each month have been calculated, half of the information needed to produce a cash flow projection is available. The other half consists of a forecast of likely receipts. This must again be based on the likely progress forecast in the bar chart and a knowledge of the contract documents which will show when interim accounts can be prepared and submitted.

Based on unit prices

In forecasting receipts the contractor will use the unit prices shown in the bill of quantities (rather than estimated ***costs***), so in this case the allowance for profit will be included.

Deduct retention

When the likely gross value of the interim accounts has been calculated, the contractor should then deduct a sum to cover the retention money that will be held by the client against satisfactory completion. This will be specified in the contract document, but is often 10% of interim valuations and 5% of the final certificate.

Date of payment

In cash flow terms, the contractor is less interested in the date that the account is submitted than in the date on which he is likely to receive a cheque from his client in settlement. This must be guessed on the basis of past experience with the client, but an interval of 2-3 weeks is quite common.

Estimated receipts

With these considerations in mind, let us suppose that the contractor has calculated that he will receive the following amounts from his client on the following dates:

Interim Account No.1 – $1500 on 20 February
Interim Account No.2 – $2500 on 20 March
Interim Account No.3 – $4000 on 20 April
Interim Account No.4 – $4500 on 20 May
Interim Account No.5 – $3500 on 20 June
Final Account – $2050 on 20 July

(leaving $950 retention money which will not be released until the end of the maintenance period).

Cash flow forecast

Now that both monthly payments and monthly receipts

have been separately forecast, we are in a position to put them together to form a cash flow forecast.

Cash flow forecast – Four houses

Month	*Payments*	*Receipts*	*Cum. Payments*	*Cum. Receipts*	*Cash Position*
January	1575	–	1575	–	–1575
February	2375	1500	3950	1500	–2450
March	3825	2500	7775	4000	–3775
April	3825	4000	11600	8000	–3600
May	3250	4500	14850	12500	–2350
June	1150	3500	16000	16000	–
July	–	2050	16000	18050	+2050

Analysis

The contractor will note that the maximum cash requirement shown from this forecast is $3775, so he would be wise to allow for about $4000 to be on the safe side. If everything on the site goes according to plan, the contract should break even in cash flow terms in June, and a surplus of $2050 should result by the end of July.

Combining contract cash flow projections

Although a contractor is interested in the cash flow projection for every contract which he undertakes, it is only by combining them that he can forecast his overall business cash needs. Unless he is very unlucky, he will find that the peak cash deficits on his various contracts will occur at different times, so that the call on working capital will even out over a period.

Example

To show the effect of combining contract cash flow projections, we will assume that the contract for four houses is the first contract to be undertaken by this contractor. We will also assume that he expects to undertake three other contracts in the current year, and that he has certain central costs which are not allocated to contracts (personal expenses, etc.) which will run at $500 per month.

Rough estimates for later contracts

Naturally the figures for the later contracts will not be based on detailed programmes in the same way as the first contract, but a rough estimate based on intended turnover will give an idea of the likely cash needs on these contracts. In fact if the cash needs are in excess of available working capital, the contractor may have to revise his intended turnover figures downwards. The reader will note that the following table is based on the cumulative position in each case, so that the central costs figures rise by $500 per month.

Cash position for current year

	Contract Number				*Central*	
	1	*2*	*3*	*4*	*Costs*	*Net*
January	–1575	– 500			– 500	–2575
February	–2450	–1500			–1000	–4950
March	–3775	– 300	– 100		–1500	–5675
April	–3600	100	– 750		–2000	–6250
May	–2350	750	– 200		–2500	–4300
June	–	750	–	– 400	–3000	–2650
July	2050	750	300	–1000	–3500	–1400
August	2050	750	1000	–2500	–4000	–2700
September	2050	750	1000	–4000	–4500	–4700
October	2050	750	1000	–1000	–5000	–2200
November	2050	750	1000	500	–5500	–1200
December	2050	750	1000	1700	–6000	– 500

Analysis

The table shows that the highest demand for working capital will be $6250 in April. If we assume that the contractor has sufficient funds to cover that demand, it is clear that there will be sufficient to cover his needs later in the year (providing the forecast is realistic). This means that the contractor could take on additional work without straining his working capital, although he would have to avoid contracts which would need more than about $1500 finance in September since there is already a secondary peak of $4700 in that month.

Regular updating

Naturally it will not be good enough to prepare a schedule of this kind in January and assume that it will then hold good for the rest of the year. Just as programmes have to be updated to take account of the way things work out in practice, so do company cash flow forecasts.

Monthly intervals

This updating should normally take place at monthly intervals. The procedure is to redraw the table with new figures for existing contract cash positions where appropriate and estimated figures for any new contracts that have been awarded since the previous month. Normally the contractor will also add a further month to the table on each occasion, so that the cash flow forecast always covers a complete year ahead.

Jobbing work

Of course some builders rely on a series of smaller jobbing activities rather than a relatively small number of larger contracts. They will approach the problem in a different way, since it would be too complicated to calculate the cash flow projection on every small job separately.

Procedure the same

The procedure however is the same. We start by asking "what do we know?". Then we go on to estimate what is likely to happen in the future on the basis of past experience.

Example

Let us imagine that we have been called in to advise a small builder who is worried about his immediate future.

The facts

These are the facts that he presents us with:

1. Cash in hand (or at bank) totals $300;
2. Customers have been billed for $3500, but have not paid yet;
3. Suppliers have sent statements totalling $1000 but have not been paid yet;
4. Taxation of $400 is due for payment.

The estimates

Now we go on from these facts to estimate what is likely to happen in the next four weeks:

1. Customers already billed will pay $2500, leaving $1000 still owing;
2. Suppliers will expect to be paid their $1000 within that time;
3. Gross wages are $400 per week;
4. The proprietor needs $30 per week for personal expenses;
5. Petty cash costs are $20 per week;
6. Materials costing $850 will be delivered, but will be on monthly credit account;

7. Jobs finishing this month are:
   i. $500
   ii. $750
   iii. $300
   iv. $1050

(none are likely to pay within this period).

The forecast

These facts and estimates allow us to forecast what is likely to be the contractor's position in four weeks' time. This is done by once again gathering together estimates of payments and receipts separately. The difference between the two will give the actual surplus (or deficit) at the end of the four week period. The forecast for this example is shown below.

Cash forecast

Four weeks to . . . .

Receipts		
From customers		2500
Payments		
Taxation	400	
Materials supplies	1000	
Wages (4 x $400)	1600	
Proprietor (4 x $30)	120	
Petty cash (4 x $20)	80	3200
Forecast deficit for 4 weeks		$ 700

Danger

Since the builder only started with $300 in cash at the beginning of the period, this forecast signals danger. It is clear that the proprietor of the firm did well to look for advice, and was well justified in his concern for his firm's immediate future!

Three possibilities

Now let us consider what we could suggest to help him out of his difficulties. If we find that his business is run reasonably efficiently and there is no scope for economies on labour or material costs, there are three possibilities:

1. Accelerate payments from customers;

2. Slow payments to creditors;
3. Approach bank manager for a loan.

Approach customers

The first possibility is to approach the firm's customers and attempt to get them to pay more rapidly for work done. We could start by approaching the customers who owe $1000 in respect of work already done, and who will be more than one month overdue in making payment by the end of the period. We know that a customer is a VIP to any contractor, but we can reasonably expect him to behave like a VIP and pay his debts on time.

Debt collecting

When attempting to collect overdue debts, the first approach should be very polite and in the form of a reminder in case the customer has overlooked the account. If that does not produce results, it will be necessary to be progressively firmer and, in the end, a lawyer may have to be brought in. If a customer has no good reason for late payments, he is not much of a VIP! If he doesn't pay at all, he is more like a thief and his continued custom is worth less than nothing.

Next four jobs

If it seems unlikely that the outstanding sum of $1000 will be collected in the next four weeks, there is another possibility of obtaining accelerated receipts. We know that four jobs worth a total of $2600 will be completed in the next four weeks. It might be possible to ask some of these clients for payments on account, if necessary offering a modest discount for early payment.

Slowing payments to creditors

A less desirable possibility is to slow down payments to the firm's creditors. In fact, if there is no money left we know that they cannot be paid. On our forecast we will be short of $400 at the end of the four week period, which would mean that $400 out of debts of $1000 could not be paid. This would mean that the contractor would get a reputation as a bad payer, and the suppliers might refuse him credit for materials needed on current contracts which would mean that a difficult cash problem would become impossible.

Poor alternative

If there is no other possibility at all, the only answer would be for the contractor to explain the problem to his suppliers and ask permission to delay payment. This is better than waiting for supplies to be cut off, but it would still leave the suppliers with a doubt about the stability of the contractor's business and his worth as a customer to them. In addition, the contractor would lose cash discounts which are very valuable in cutting costs.

Approach bank manager

The final possibility is for the contractor to approach his bank manager and ask him to make overdraft facilities available for the next few weeks. If the contractor has been a good customer of the bank, he may agree to allow the contractor to overdraw up to $500 which should just allow the contractor to survive the next four weeks. It will be necessary to extend the cash flow projection to show when the cash flow position will soon improve so that the overdraft can be repaid.

Bank overdrafts

Bank overdrafts, if they can be obtained, are a good way of dealing with temporary cash deficits since interest is only paid while the account is overdrawn and the interest is calculated on the balance outstanding at the time. For example, an ordinary loan of $1000 for a year at 8% interest would cost $80 per annum. But, if the deficit will probably only last for three months, an overdraft at 12% would be better since interest would be related to the amount overdrawn in any month. If the average amounts are $800, $1000 and $500 the cost would be:

$$800 \times \frac{(12)}{(100)} \times \frac{(1)}{(12)} = 8$$

$$1000 \times \frac{(12)}{(100)} \times \frac{(1)}{(12)} = 10$$

$$500 \times \frac{(12)}{(100)} \times \frac{(1)}{(12)} = 5$$

Total $23

Thus, in this case, an overdraft at 12% would cost just over a quarter of a loan at 8%.

Chapter Nine

Monitoring Progress

Monitoring progress and checking profit and cash flow against budget. Variance analysis. Principles of work study. Productivity and incentives.

Need to know in time

Earlier we have discussed management tools such as planning, programming, budgeting and cash flow analysis. These are all very valuable tools in the contractor's battle for survival and progress. But the preparation of all these plans will not of itself assure success. We need to know whether everything is going according to plan. If not (and the average building business operates in such an uncertain and risky climate that something is almost certain to miss its target) we need to know exactly what is wrong in time to do something about it.

Control

The good manager is in ***control*** of his business. Being in control does not just mean having the biggest office and driving the biggest car. It means knowing what is going right and what is going wrong. It also means taking appropriate action as soon as it is needed so that difficulties are prevented from turning into disasters and opportunities are grabbed while they are available.

Planning — Information — Action

There are three aspects to control:

- planning is the ***basis*** for control;
- information is the ***guide*** for the manager;
- action must be taken by the manager after analysing the plan and the information.

Every employee must control own activities

Although control starts at the top, it will only be successful if it works down the line so that every employee is in control of his own activities. The site agent must know the programme for completion of his job and must gather sufficient information on output, delivery of materials, etc. to be in control of work on the site. His carpentry foreman must know how many doors and windows he is required to hang in the next week, and be in a position to assess whether the target will be met. Then the carpenter must be given the information, tools and materials to get on with his job without difficulty or delay.

Information must flow upwards

Just as control works down the line to the man on the site, information must come up the line to the manager of the business. He needs this information to check profit, cash flow and other key figures regularly against the budget.

Checking variations

If the plan or budget has been prepared on a realistic basis, the actual figures achieved on the site should not be very different from it. Thus a good deal of time and effort can be saved by limiting the information required to variations from the plan or budget. This will tend to highlight problem areas, such as contracts which are running late or cost headings which are higher than anticipated.

Management by exception

The principle here is one of 'management by exception'. Time is always short for a manager, and his job is to sort out problems. By looking mainly at the variations from the budget or plan, he will find that the problems present themselves for analysis – so at least he will not have to waste time looking for them.

Cost control

Once a contract has been signed, the selling price of his product is fixed as far as the contractor is concerned. This may be either in terms of an agreed lump sum broken down into stage payments or as unit rates in a bill of quantities. Thus he does not have the same freedom that a shopkeeper enjoys to lower or raise prices as costs change. For a contractor the only way to achieve a better profit on a contract is to cut costs, so cost control on the site has to be one of the key areas of concern for any contractor.

Four cost headings

All costs come under one of the following four headings,

and each makes its own demands on scarce working capital:

1. Labour;
2. Materials;
3. Plant, vehicles and tools;
4. Overheads.

Labour

Labour costs consist of wages and associated overheads such as insurance, holiday pay, sick leave and taxation. It is useful to keep a regular check on the number of man hours that are required to complete various standard tasks such as excavation, concreting or laying blockwork or brickwork. If the number of man hours per unit of production rises, this is probably a sign that supervision is becoming less effective and general labour costs are rising.

Helps in estimating

In addition, a regular review of unit labour costs for various tasks is very helpful in ensuring that estimates for new work are based on data which is up to date.

Materials

Control of the level of stocks of materials is vital if working capital is short. Materials only start to pay for themselves at the time they are incorporated in the structure. Until then, they sit in the store or on the site needing money to finance their cost, risking loss through pilferage and adding to their effective costs every time labour is employed in moving them from one place to another. Of course it may well make sense to buy materials in advance if they are offered very cheaply or if some particular material, such as cement, is in short supply.

Control of stocks

Close attention to stock levels can also be an effective way to improve profit performance. In many contracting firms the level of profit is a fairly constant proportion of turnover, say 5%. If this is the case, the only way to increase profits without raising fresh finance is to find a way to make working capital circulate faster. If working capital at present is $10000 and turnover is $15000 due to the high level of stocks, profit at 5% is $750. But if stocks could be cut to a lower average level, a greater proportion of the $10000 could go to finance other costs and additional work could be taken on. It might even be possible to double the work done in a year to $30000 so that profit at 5% on turnover would yield $1500. Thus profit as a percentage of working capital employed in the business would rise from a below average 7½% to a more acceptable 15%.

Debtors

Another way to increase the rate of circulation of working capital is to reduce the size of the debtors figure in the balance sheet by ensuring that customers and clients pay their bills on time, and discreet pressure should be applied as soon as an account becomes overdue.

Plant and overheads

The wise construction manager will also keep a careful watch on funds tied up in plant, vehicles and tools to make sure they pay for their keep. If items are used only rarely it might be better to sell them and hire occasionally when the need arises. Overheads should also be regularly reviewed to see if they really contribute to the efficiency of the business. But in addition to these financial controls, site management must keep up a constant war on waste.

Reducing waste of construction resources

Building and civil engineering firms prosper or otherwise according to their ability and competence in using and deploying the four construction resources. These resources are:

1. Men
2. Materials
3. Machinery (plant and equipment)
4. Money

The first 'M'

Work study is concerned with the efficient use of the first of these 'Ms' — the men who are employed on the construction sites. As in the case of the other construction management techniques introduced in this book, the object is to reduce waste of resources to a minimum, so that the overall cost of building can be brought down.

Bad for workers?

It is sometimes thought that the introduction of work study techniques is bad for workers in the construction industry, as it reduces the number of man weeks required to complete jobs and thereby increases unemployment. This would seem to be in contrast with the intermediate technology approach, which emphasises labour-intensive rather than capital-intensive methods, with the object of reducing unemployment in developing countries.

Helps productivity

In fact work study and other labour-scheduling and measurement techniques are a useful part of this intermediate technology approach, as they increase the competitiveness of labour-intensive methods by increasing their productivity. Very often mechanical methods are adopted by firms and public bodies to avoid the complication of dealing with large armies of men, with the attendant man-management and supervisory problems that they involve.

Makes output measurable

They feel that the costs of using plant and machinery are easier to estimate, measure and control. Work study techniques, by making the output of individuals and groups of men ***measurable***, can increase the confidence of management when carrying out projects by labour-intensive methods.

Helps the indigenous businessman

These techniques can also aid the indigenous businessman in his competition for contracts with expatriate firms. Expatriates usually employ capital-intensive methods for two reasons. Firstly, they usually have more capital to finance their operations. Secondly, their use of expatriate staff leads to the application of a higher labour oncost to cover the higher costs of supervision and managerial resources. By employing ways of controlling and trimming labour costs, indigenous businessmen can be helped to capture a larger share of the available contracts.

Helps the employee

Thus, although the man-weeks required to complete an individual job can be reduced, these techniques could lead to a greater overall volume of work being made available. This must assist the overall prospects of local employees of the construction industry.

Increases earnings

Another advantage for employees is that the savings in costs resulting from the introduction of work study schemes allow the employer to pay an incentive bonus. These bonuses increase the earnings of the craftsmen and labourers who manage to achieve the required levels of productivity.

Written procedures

An essential aspect of good construction management is the requirement that written procedures be adopted in order to achieve management control. These procedures are sometimes felt to be unnecessary by proprietors of small building firms, but they become essential as the firms grow in size and become more complex and difficult to control.

Simple but formal

Earlier chapters have dealt with planning, programming, accountancy and budgetary control. It was shown that these simple but formal written procedures can aid the managers and directors of a business in keeping track of what is happening on their various sites.

Targets

Although hour-to-hour and usually day-to-day management decisions must be made by the agent or foreman on the site, it is important that his decisions should be made within the framework of a clearly defined set of written expectations or targets.

Range of expectations

The reason for providing written targets for output and performance is to provide standards against which actual results can be measured. In the simplest form, a small gang may be sent off to complete a small repair or maintenance job and be told they have three days in which to finish it.

Top of range

If they complete in two days, they know that their boss will be pleased and may well award them a bonus payment. Performance at this level would represent the top point of the range of expectations.

Centre of range

If they complete the work in the three days allotted to them they know that he will at least be satisfied with their performance. This performance would lie at the centre of the range of expectations.

Unsatisfactory performance

Performance below the target set naturally represents unsatisfactory performance. If the job takes four days, they know that they will have to think up some excuse to explain away the extra time, such as the work proving more complicated than expected, difficulty in obtaining materials or tools or difficulty in satisfying the client or Clerk of Works. If it takes more than four days, they know that they will probably be dismissed or transferred to lower-paid work. This represents the lower end of a typical (but simplified) 'range of expectations'.

Management to be aware

However, for this 'range of expectations' to apply, they must know that their work is being ***watched*** and that their manager will be ***aware*** of the output that is achieved. The need for management to be aware in this way is, of course, even greater in large firms where directors and top management can only know a small proportion of their employees by name.

Operations fully exposed

The Finance Director of one of the largest and most successful electrical engineering businesses in the UK has said that:

> "We must make people feel that somebody who really knows is watching them all the time. They must feel that their operations are fully exposed."

Somebody who really knows

Of course, the key phrase is 'somebody who really knows'. It is easy for the man in charge of a firm with 10 employees to really know how well they are doing. It is more difficult when the workforce has grown to 50. When it reaches 100 he has to rely on reports from his foremen or charge hands. It is at this stage that many small businesses get completely out of control, as the informal methods which have brought success in the past become inadequate to cope with the sheer value of output and numbers of men employed. The introduction of the disciplines and techniques of work study is one way for the management of these medium/large construction firms to ***really know*** the productivity and output of the men they employ.

Setting standards

Work study can be seen as a way of ***setting standards*** in the difficult field of human effort and activity. We depend greatly on the measures of physical properties, such as weight, volume and distance, to describe the extent of civil engineering and building contracts shown on the drawings and in the bill of quantities and specification. These figures set ***standards*** for the contractor measuring the quantity, and therefore the ***cost***, of the materials he will need to complete the job.

Time standards

It is obvious that there would be great value in trying to set usable standards to express the amount of time, and therefore the ***cost***, required for craftsmen and labourers to complete specified tasks. These standards would bring more system into estimating, programming and site management as well as providing a factual basis for the introduction of incentive and bonus schemes.

Definition

A simple definition of ***work measurement*** could be:

The application of techniques designed to establish the time for a ***qualified worker*** to carry out a ***specified*** job at a ***defined level of performance.***

A more detailed definition of work study generally is given in the British Standard Glossary of Terms B.S. 3.38 as follows:

A management service based on those techniques, particularly ***Method Study*** and ***Work Measurement***, which are used in the examination of human work in all its contexts, and which lead to the systematic investigation of all the resources and factors which affect the efficiency and economy of the situation being reviewed, in order to effect improvement.

Finding a better way

In short, work study is concerned with 'finding a better way' of doing things. It is not a substitute for good management, but provides a means of studying and improving the output of a contractor's staff.

Two techniques

Work study techniques can be divided into two main areas:

1. Method study
2. Work measurement (time study).

Method study

Method study looks into activities and asks questions about them:

Why do we do this job?
don't we do something else?
What is the result of what we are doing?
else could we do?
Where is the work done?
could it be done more effectively?
How is it done?
else could it be done?
When is it done?
could it be done?
should it be done?
Who does the work?
else could do the work?
should do the work?

Critical examination

In fact we could say that the method study consists of a critical examination of an activity in order to determine the most economical and effective means of getting things done.

Work measurement

Work measurement (time study) is a way of measuring

how much time is required to complete any given activity or job. Since different people work at different rates, the performance of the individuals or teams involved must be taken into account. Thus work measurement helps to take the guesswork out of estimating labour costs.

Useful to public sector

These techniques are just as useful in the public sector as to the owners and managers of private firms and companies. Ministries of Works and engineers and architects in public service are required to keep building costs within their estimates on direct labour as well as on contracts put out for tender. Work study techniques can help them to control and reduce the impact of high labour costs and improve productivity.

Not just for manual workers

Many managers assume that work study techniques can only be used to improve productivity among manual labourers. This is not correct, because similar techniques (sometimes described as O&M — organisation and method study) can be applied to clerical workers, supervisors and even managers themselves. In fact, by proper recording of the way he uses his time in his office diary, the manager himself can build up a useful record of his own output.

The manager's output

Whether or not your firm uses work study, you, as a

manager, should be aware of your work content and output. If your diary records are properly kept, the time taken up by various activities over the course of a month can be analysed. A typical time allocation for the Contracts Manager of a small/medium size building contracting business is given below.

CONTRACTS MANAGER	HOURS/MONTH
ANALYSING SITE REPORTS AND OPERATING FIGURES	50
CONTACTS WITH SUPPLIERS	20
" " CLIENTS	20
EXAMINING + PREPARING TENDERS	20
BOARD + COMMITTEE MEETINGS	8
CONTACTS WITH SUBORDINATES	18
TRAVELLING BETWEEN OFFICE + SITE	16
SITE INSPECTIONS	20
CONTACTS WITH SPECIALISTS (ACCOUNTANTS, ESTIMATORS ETC.)	20
CONSIDERING FUTURE POLICY	8
	200

No ideal time allocation

These are, of course, only typical figures and by no means an ideal for every manager to try to copy. In fact the idea varies from firm to firm, from job to job and even from individual to individual. The important thing is that every manager should be aware of what he spends time on and perhaps even more important, what he does ***not*** spend time on.

Policy making

The general principle is that the higher a manager's job in his organisation's hierarchy, or 'family tree', the more time he should devote to general policy making and the less time he should give to routine administration.

What and how to delegate

The good manager should know ***what*** to delegate and ***how*** to delegate, and this means understanding the strengths and weaknesses of all his immediate subordinates. Unfortunately, the temptation, and the weakness, of many managers is to delegate only the jobs that they dislike and to hang on to the work that they understand and enjoy.

Neglect

Thus the manager who started as an accountant may do the book-keeping himself, but will tend to visit his sites only rarely and will leave negotiations with clients and suppliers to others. The manager who started as a foreman or site agent will bias his time in the opposite direction, and may neglect important cost accounting and control procedures.

Developing 'all round managers'

Often this neglect by managers of responsibilities outside their own specialist training is a result of lack of knowledge, and therefore lack of confidence, in these 'foreign' areas of expertise. By attending appropriate training courses in these management subjects, they can be helped to develop into 'all round managers' who can confidently tackle the many and varied problems faced by building contractors.

Time means money

Although it is a common saying that 'time means money', many contractors (and even more of their employees) fail to realise the cost of their time on the job, and the losses which will result if this time is not used effectively.

Using time effectively

In fact, it is by saving time on your jobs and ***using time effectively*** that the largest cost savings can be made. For most contractors, the basic cost of a bag of cement or a length of timber is roughly the same. So are the costs of buying plant, tools and equipment. So are the basic hourly and daily costs of employing masons, carpenters and labourers.

The cost equation

In fact we could draw up a cost equation, which would apply to every construction job. The materials cost would be fixed, but the other costs generally depend on how long the job takes. In the illustration we can see a general cost equation.

Cost equation example

For example, this general cost equation could be applied to a particular job, possibly building a small factory, in order to demonstrate how it works out in practice.

Assumptions

Suppose materials cost \$20000

Weekly labour cost (av.) = \$200

Weekly plant cost = $ 40
Weekly overhead cost = $ 60

Estimated time for completion = 40 weeks

THE COST EQUATION:

MATERIALS COST + [(LABOUR WEEKLY COST) + (PLANT WEEKLY COST) + (OVERHEAD WEEKLY COST)] x TIME

= TOTAL COST

Labour force

To simplify the example we will assume that all the men are on the site for the full period of 40 weeks, although in practice the labour force would vary as different trades deal with different parts of the job.

Variable costs $300

The weekly costs ('variable costs') total $300, so the tender price (allowing a 10% profit margin) would be built up as shown below.

THE COST EQUATION:

MATERIALS	20 000
VARIABLE COSTS: $300 x 40 weeks	12 000
TOTAL COST	32 000
ALLOW 10% PROFIT	3 200
TENDER PRICE	35 200

If they get the job

Suppose their tender is successful. Now they have the problem of completing the job at a minimum cost, so they must examine the figures they used in preparing the estimate and try to find some savings.

Looking for savings

They will look for savings on purchasing of materials, but it is unlikely (providing the estimate was realistic) that very large price variations between suppliers will be seen, so the actual cost will probably be quite close to the estimate. Another source of possible savings is by cutting down on wastage and stealing of materials from the site. This would probably be reduced if the time taken to complete the job could be reduced.

Cutting time to do the job

The weekly cost of labour, plant and overheads (office staff, etc.) is also likely to be in line with the estimate, so the only hope of a substantial saving is by encouraging the men to work more quickly and cutting the ***time*** taken to do the job. If it would encourage and ***motivate*** them to work more effectively, it would even be worth while giving them an incentive bonus to complete the job ahead of target.

Incentive schemes must be fair

But if an incentive scheme is to be successful, it must be seen to be ***fair***. It is no use waiting until the end of the job, and then giving bonuses to whoever ***seems*** to have earned them. This just causes resentment among the other workers, and is not even a real incentive scheme since the men do not know in advance what would be their reward for working more effectively.

Productivity increases

It has become generally accepted that a reasonably generous incentive scheme can increase productivity by about one-third. Thus a blocklayer who has been laying 300 blocks in a day could be expected to lay 400 under an incentive scheme, and the general labourer who had been excavating 6 cubic metres could be expected to excavate 8 cubic metres.

Setting ratings

This leads to the idea of setting a scale of 'ratings' which would allow the manager to ***measure*** the output of workers in different trades, instead of guessing or saying 'my masons are better than most in this town, but my carpenters are below average'.

The B.S. rating scale

There are several rating scales, but one of the most commonly used is that put forward in British Standard BS 3138 where 0 represents no activity and 100 represents 'standard rating' or the performance of a task in a brisk and business-like way, compatible with a trained worker motivated by the prospect of an appreciable bonus if agreed targets are reached.

'75' rating

There is of course a full range of intermediate levels of performance, the most important of which is 75 (three-quarters of standard rating) which is the steady, deliberate, unhurried performance of a worker under proper supervision, but without any kind of financial incentive beyond his hourly or daily basic wage rate.

The illustration emphasises some of the different levels of performance:

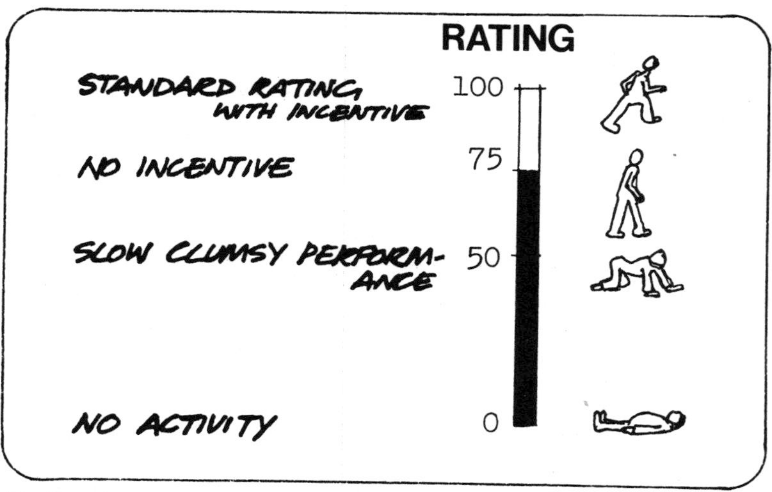

Estimating ratings

You will note that ratings can be applied to any kind of work, from excavation to fixing roof sheets. The drawback would seem to be the difficulty of judging the actual ratings of particular workers on particular jobs. In fact, with a little practice, it is possible to get reasonable agreement between different observers on the ratings to be applied to various performances. Thus this objection is more theoretical than practical, and ratings can be quite objective although the process is very time-consuming for a small contractor.

Rating over '100'

Although 100 represents 'standard rating', performance levels in excess of 100 are, of course, quite possible. However, they are unlikely to be achieved unless the operator is super-skilled and very highly motivated.

Back to example

Returning to our example of the cost equation, we expected the job to take 40 weeks if no bonus incentive payments were made to the employees. In other words, we expected completion in 40 weeks at a performance rating of 75. However, if we were to decide to start an incentive scheme which brought the performance rating up to 100, it would mean that an extra 33⅓% work would be completed in every day and in every week. The progress of the job week by week would be as shown in the illustration.

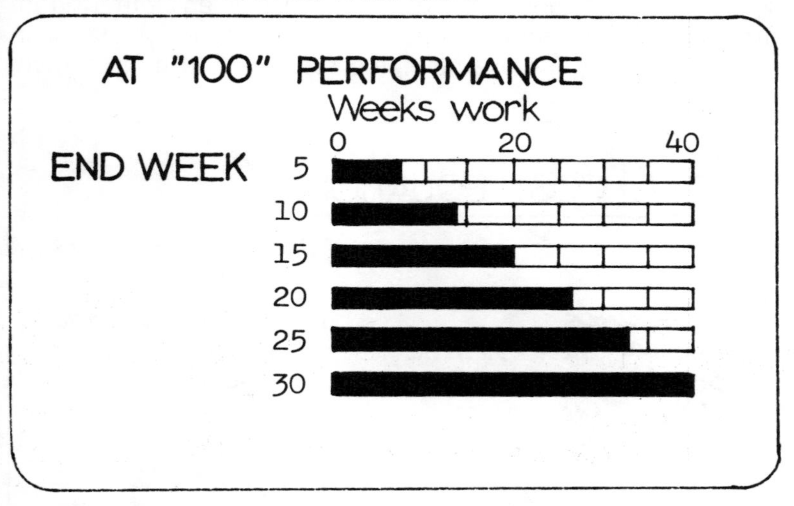

So: at the end of week 1, 1⅓ weeks' work would have been done,

at the end of week 2, 2⅔ weeks' work would have been done,

at the end of week 3, 4 weeks' work would have been done,

and so on until the job would be complete at the end of week 30.

Saving ten weeks

The result of installing the incentive scheme is that, although we had estimated to have required 40 weeks' labour, plant and overhead costs, the actual expenditure would amount to only 30 weeks' costs. Thus the gross saving would seem to be 10 weeks' labour, plant and overhead costs.

Calculating the bonus

However, we must allow for the cost of the incentive bonus to be paid to the employees if their performance reaches a 100 rating. Usually the bonus is paid at the rate of one-third of the employee's basic wage; in other words, his basic wage goes up by one-third if his output goes up by one-third.

Useful savings

At first sight it may seem that the employer is giving away all he has gained, but in fact he will be making a useful saving on plant and general overhead expenses, together with a small saving on labour oncosts and wastage and pilfering of materials. If we apply a one-third bonus rate for standard performance to our example, the result would be as shown below.

THE COST EQUATION

	ACTUAL	ESTIMATE
MATERIALS	20 000	20 000
LABOUR $200 × 30 WEEKS	6 000	8 000
BONUS ONE THIRD	2 000	–
PLANT $40 × 30 WEEKS	1 200	1 600
OVERHEADS $60 × 30 WEEKS	1 800	2 400
	31 000	32 000
PROFIT	4 200	3 200
TENDER PRICE	35 200	35 200

Profit margin increased

Despite the payment of a bonus totalling $2000, the total cost of carrying out the work has been reduced by $1000 giving an additional profit of $1000. In fact the profit margin has increased from 10% to just over 13½%.

Typical result

Although this example has been greatly simplified, it does

demonstrate the sort of result that a contractor could hope to achieve when he introduces a well thought-out incentive scheme. Both he and his employees should gain from the better productivity that can be achieved by a more purposeful method of working.

Chapter Ten

Using Work Study Techniques

Measuring and controlling

In the previous chapter the advantages of work study to both the employer and his staff were discussed, and an example showed how it is possible to pay a bonus to site operatives if they achieve increased output and thus show an improved profit for the owners of the firm. These improvements result from measuring and controlling labour output in a rational and scientific way.

Finding a better way

This final chapter is devoted to the 'how' and 'when' of work study. It is certainly not suggested that all the techniques that are mentioned can or should be applied in the small building enterprise. The principles of work study are important, however, as they lead to systematic thought about working methods. The successful contractor should always be thinking about different ways of organising his activities and operations on the site, and the author hopes that this chapter might help him to reach his goal of 'finding a better way'.

Two groups of techniques

We saw that work study has two separate groups of techniques designed to meet two separate general objectives:

Method study — To compare ways of getting a job done and choose the method that requires the lowest effort and employs least resources.

Work measurement — Once the method of carrying out a job has been chosen, work measurement (or time study) can be used to give a standard time to complete it.

Method study

Method study is a step-by-step procedure for improving the output of a private firm or of a government direct labour force. It is often possible to introduce methods of working that reduce expensive double handling of materials, or ensure that dumpers and concrete mixers are fully used instead of being used only 10 or 15 minutes in an hour. Very often these improved methods are not difficult to devise and carry out, but can lead to quite large savings of time and effort and thus reduce costs.

Step-by-step

I said that method study is a step-by-step procedure. It is a systematic way of looking at a process, or a series of individual operations, to see which of them can be simplified, rearranged, combined or even left out altogether. A typical example of a process made up of a series of individual operations is the manufacture on site of concrete blocks, as shown in the illustration.

Study

By using method study techniques, it is possible to study each operation systematically, and ensure that the labourers are all fully employed and that the machine is kept going all the time. This means that the cost of each hundred blocks will be kept down, and the job should become more profitable.

1. Select the work to be studied

Select the work to be studied. Since the small contractor has only limited time and resources, it will not be possible to work study every job from hammering in the pegs to set out the building to sweeping up on completion. He has to ***choose*** which operations would be most likely to show big savings from improved methods. This choice will depend on his judgement as a manager, but some jobs will almost 'choose themselves' due to:

— Costs higher than estimated.
— Bottlenecks in production which slow down other operations.
— Labour or plant standing idle for long periods.
— Excessive handling of materials.
— Excessive scrap or rejected work.
— Bad working conditions.

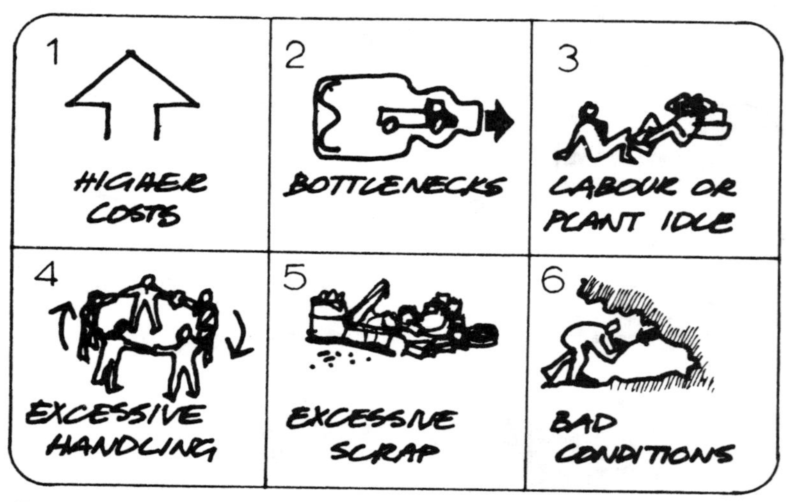

Overall quantity and value

Of course the overall quantity and value of the work to be studied must be considered. There is no point in carrying out a careful study of trench excavation and pipe laying if there are only a few drains to be laid on the site. The job to be studied should be one which represents a high proportion of the overall cost of the project and which, from the past experience of the manager, could be carried out more effectively as a result of systematic study.

2. Observe procedures

Study carefully the way in which the work is being carried out. This can be done by directly watching the operation, but it may also be necessary to study specifications, bills of quantities and drawings, and other records such as time sheets and plant usage sheets. In advanced applications, work study practitioners sometimes even use photography to get a clearer idea of the work and movements involved.

Experience

This step must be carried out by a well-trained and experienced observer, who will need an orderly and questioning mind. He will also need an ability to get on with people, so that he can discuss the operation with everyone concerned to ensure that he obtains facts rather than opinions or prejudices. He will try to find out whether, as is often the case in building work, the operation that takes place on the site is very different from that anticipated by the estimator who priced the job. It may be carried out by a different number of men, in different conditions, in a different place and by different methods.

A better way

This does not necessarily mean that the men on the site are wrong. They may simply have found a better way of doing the job by practical experience. But if this is the case it should be reported back, so that the improved method can be used on other jobs and future estimates can be made more competitive by allowing for the method that will actually be used.

3. Record

Write down all the relevant facts on the methods that are being used. These records may well include charts and diagrams which make the operation easier to understand. Many special forms of recording aid have been developed to suit particular conditions and requirements, although these will normally only be used by specialist work study experts.

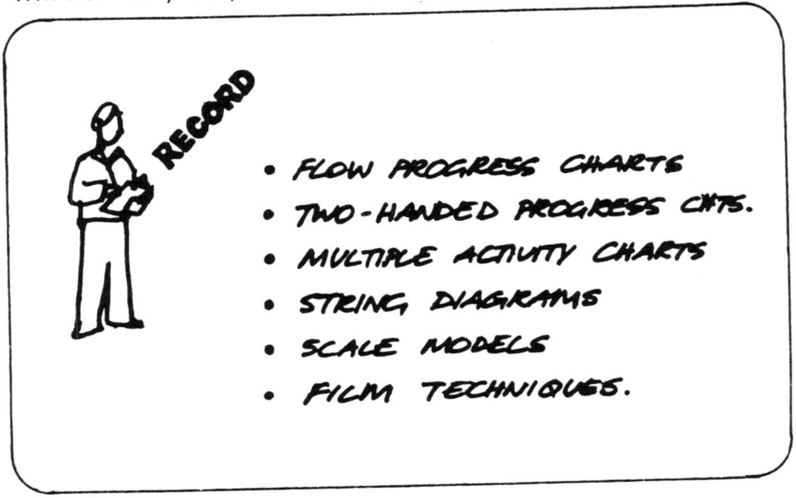

Flow Process Charts for recording activities of workers, processes or movement of materials. Special symbols indicate, at each stage, whether the item consists of a manufacturing, transport, inspection or storage operation.

Two-Handed Process Charts for showing movements of the operators body members, which may lead to simplification of operations.

Multiple Activity Charts for recording the simultaneous movements of man and machines.

String Diagrams where string is used to indicate the paths taken by men and materials in carrying out a job. On work such as the bending and fixing of reinforcing rods this can lead to considerable savings by relocating materials and equipment to cut out unnecessary movements.

Scale Models to assist in planning production shops.

Various Film Techniques including memo motion films, which can be speeded up to show 8 hours' activities in 20 or 30 minutes.

4. Examine

Examine systematically the facts about the activity. The third step should have provided the present facts about the various stages of the operation:

Its Purpose:	What is achieved?
Its Place:	Where is it done?
The Sequence:	When are the various stages done?
The Person:	Who does it?
The Method:	How is it done?

10.5

- PURPOSE
- PLACE
- SEQUENCE
- PERSON
- METHOD

Challenge the facts

With these answers available, we should now ***challenge*** these facts by asking more questions:

The Purpose:	Why is it done? Is it really necessary?
The Place:	Why is it done there?
The Sequence:	Why is it done in that order?
The Person:	Why that person? Could someone less skilled do it?
The Method:	Why is it done in that way?

Some answers are obvious

Of course some of the answers will be obvious. We know why a blocklayer builds a wall. We know that the wall has to be built where it is shown on the plan. We know that a blocklayer has to do the job. But it may be possible to challenge the sequence and the method.

Breaking down complex operations

In other operations, all five points could be challenged. Sometimes, on a large site, a complicated operation which requires skilled men can be broken down into a series of simple tasks which could be carried out by unskilled labourers.

5. Develop improved methods

Having completed the criticism of the present method, the next set of questions should lead us to other methods which might be more satisfactory:

Other

Other Purposes:	What else could be done?
Other Places:	Where else could it be done?
Other Sequences:	Could the stages of the operation be carried out in a different order?
Other Persons:	Who else could do it?
Other Methods:	How else could it be done: Would different machinery or equipment help?

Most attractive alternative

Out of these alternatives, one may appear to be more attractive than the others. In most cases, this will be the method which shows the lowest cost, but it may be the method which takes least time or interferes least with other operations on the site.

Hidden snags?

When the most favourable method has been chosen, it should be subjected to critical examination to ensure that there are no hidden snags:

Its Purpose:	What is achieved?
Its Place:	Where should it be done?
Its Sequence:	Are the stages of the operation carried out in the best order?
The Person:	Who should do it?
The Method:	How should it be done?

Types of solution

The type of solution to be adopted will depend partly on the nature of the operation and whether the suggested change would lead to additional capital investment in plant and equipment or heavy training costs. This may be the case with improved methods in a joinery shop. Although new woodworking machinery would lead to higher productivity and apparently lower costs, the cost of scrapping existing machinery and building an extension to accommodate the new plant may put too great a strain on the firm's capital resources. In a case like this, it might be better to plan a series of phased improvements.

6. Install

The sixth step is to install (or introduce) the improved method. It is important that careful thought and consideration be given to this step, as a good method can fail due to a poor flow of materials, inadequate tools or untrained operatives.

Persuade

It must be remembered that the operatives themselves must be ***persuaded*** of the value of the new method. If they are simply ***ordered*** to do their work in a different way, they will go about it half-heartedly and almost certainly find some reason to make it appear more difficult than the well-tried methods they are used to. It is for this reason that it is vital that the work study officer, besides being technically competent, should be able to get on well with the people concerned and explain in simple language how the improved method will make their jobs easier and more productive.

7. Maintain

The seventh step is a continuing process of maintaining the improved method. This will involve periodic checks on the new method to ensure that it continues to operate as it was planned to do. It may also be necessary to allow for changes in external conditions, new staff or equipment, and it may be that further changes which would further improve efficiency will become obvious when the new method is in operation.

Review

Thus there are seven steps on the 'method study path' to an improved procedure for an operation, as shown in the illustration:

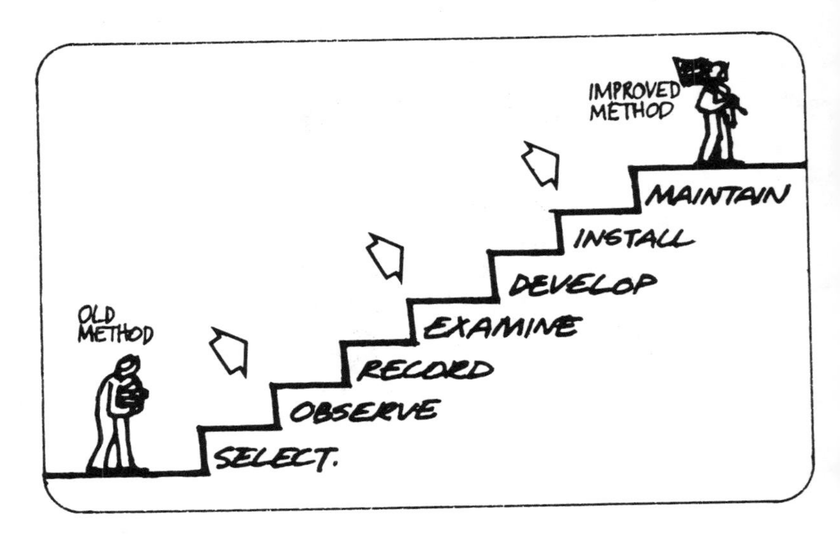

Work measurement

The other major aspect of work study is of course ***work measurement*** or ***time study***. Work measurement is an important tool of construction management in that it offers the possibility of a rational system for improving management control over labour costs and increasing the output and productivity of building workers. It offers the manager an opportunity to ***measure*** the output of his men in the same way as he measures the number of bags of cement and truck loads of sand and aggregate required to complete a building project.

Greater earnings

Besides helping management in the estimating, planning and control of work, we saw in the previous chapter that it could lead to greater earnings by the introduction of a bonus incentive scheme.

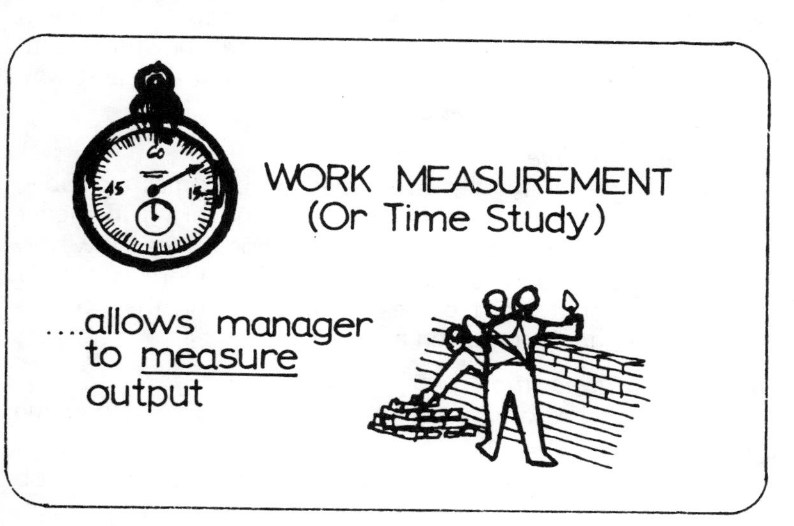

How to measure work

In this section, we will be looking at the 'how' of work measurement — how we go about setting standard times for the many and varied jobs that are carried out on a building site. We will also discuss how we can use these standard times in the day to day control of operations in a contracting firm or a Public Works Department.

Getting the method right

The first step is of course to select the work to be measured. Before measuring a particular task, it is important to be sure that the ***method*** used to carry out the work is right. If the operator is timed for carrying out a particular job, such as building a concrete block wall or hanging doors, he will be given a target based on the method used. If he later finds a better way of doing the job so that he can meet the target productivity with less effort, he will expect the previous level of bonus to continue. But his fellow workers will still be struggling to meet their targets for other jobs, and they may ask for their targets to be reduced so that everyone has to work at the same pace.

Bad to increase targets

However, if the management tries to increase the target

of the man who has found a better method, he will be very unhappy and will not co-operate again by trying to improve methods and performance. The only answer to this problem is to make sure, if necessary by applying method study techniques, that the job is being done in the most efficient way possible ***before time studies are started.***

Re-timing is possible

Where a genuinely improved method becomes available, it is of course possible to re-time jobs and bring in higher output targets or lower target times for operatives. However, re-timing should only be carried out when there is a clear justification for it, as it causes resentment and the workers feel that they are being taken advantage of.

Other uses of time study

Time study has other uses than as a basis for introducing bonus incentive schemes.

1. It can be used to gain factual information on the job contents before carrying out a method study.
2. Time studies on two methods of carrying out the same job can be useful in deciding which one is more productive.
3. Where plant or machinery does not seem to be producing as much work as expected, a time study can help management to find out what is wrong.
4. If the cost of making or building certain items, such as window frames or concrete blocks, is higher than estimated, a time study can help management to work out a better method and reduce costs.

Explaining time study

Workers everywhere are very wary of the man with the stopwatch. They feel that he is spying on them and, unless his reasons for doing this work are carefully explained, they will be resentful and uncooperative, even to the point of strike action. Thus, as pointed out in the section on method study, the work study officer needs to be human and tactful, something of a 'diplomat', as well as being well-trained and good at his job.

Introducing the idea

If time study has not been tried before in the firm or department, it is usually best to start by calling a meeting with the supervisors and representatives of the workers. At this meeting, the work study officer will explain that he has a job to do and that he wants to be fair, and will always be ready to explain ***why*** he is taking measurements and ***what*** he is writing down. It is usually best to explain the time study form he will be using, as some workers will fear that he is writing down personal comments on their work, rather than carrying out a process of scientific measurement. He will, of course, answer any questions and will probably let them handle the stopwatch themselves to show that it has no 'magical' powers.

Choosing the subject for study

Some thought must be given to which of the workers should be the subject of the time study. If possible, this choice should be left to the workers themselves, so long as they choose someone who is competent and works at least as well as the average man on the site.

Steady and unruffled

It is also important that he should be steady and unruffled as the study is made, as some workers get worried when they are watched and are unable to work at their usual pace. If the job to be studied is an important or basic one, which will later be carried out on a large scale or by several gangs of workers, it is usually best to carry out a series of studies on different subjects, and take the average as the basis for the incentive target.

New methods and new workers

If a new way of carrying out a particular job has been introduced, the time study should not be taken until the operator has been carrying it out for long enough to get really used to it. The same consideration of course applies if a new man is put into a job, as he will take time to get used to it. We all take time to learn a new skill, and our performance increases according to a 'learning curve' as illustrated.

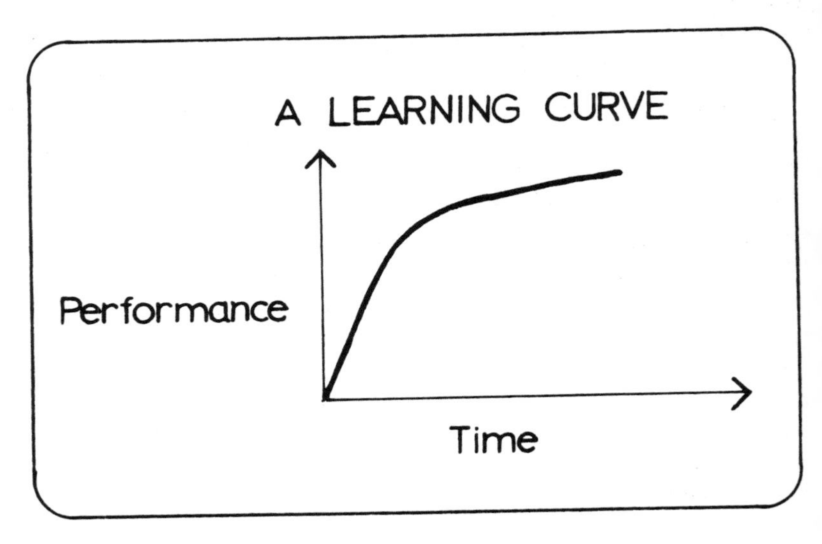

The learning curve

This 'learning curve' applies to almost every human activity, from learning to read and write to driving a car or operating a concrete mixer. At first our performance is clumsy and bad. Then we begin to 'get the hang of it' and our performance increases very rapidly. Then, as we get close to our peak performance, it begins to level off and there are very few 'tricks of the trade' left to learn which could improve our efficiency still further.

Time varies with task

Of course the time taken to reach a satisfactory level of performance depends on the difficulty of the task. We could teach a man how to check the tyre pressure on a car much more quickly than we could teach him to drive it. A man can learn to use a shovel to load a concrete mixer very quickly, but it will take several months before he reaches a satisfactory performance as a blocklayer. For the work study officer, the important consideration is that the operator to be studied must at least have been doing the job long enough to have reached the stage on the learning curve where his performance will not automatically increase as he gets more and more used to his task.

Breakdown into elements

When the operator is used to his task, the work study officer can carry out his time study. But before he reaches for his stopwatch, he must break down the job which he is going to time into elements. This is done partly to increase the accuracy of his timing, but also to separate work carried out by plant and machinery from that carried out by labourers. An example of the way a job can be broken down into elements so that a time study can be carried out is given in the illustration on the next page. This shows the nine elements, or parts of the job, that must be carried out by a gang of blockmakers.

The actual time study

The actual time study is made with the use of a stopwatch, and the work study officer records a series of times and estimated ratings for each element of the task to be measured. These figures are usually recorded on a printed time study observation sheet, as a large number of individual observations have to be made during each time study. From these observations, the work study officer is able to calculate the ***basic time*** for each element. This is done by multiplying the observed time by the percentage rating as shown in the illustration:

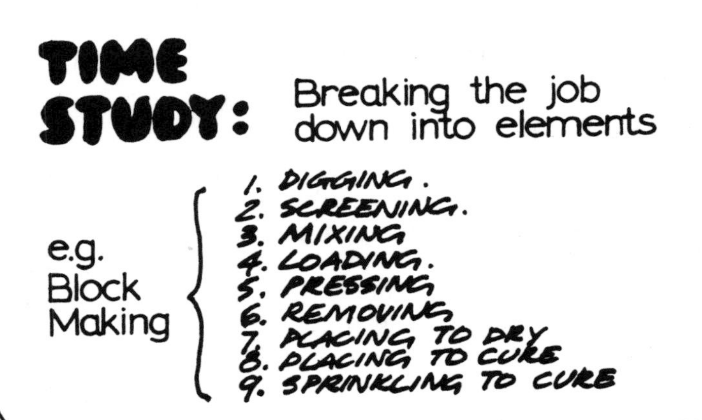

Observed time (secs)	Estimated rating	Basic time (secs)
50	50	25
30	90	27
20	120	24
26	100	26
	Average Basic time:	25.5

The illustration

This illustration represents four observations of one element of a task. Of course, the work study officer would normally take more than four observations and, since he would have chosen an experienced and steady worker, the estimated ratings would probably be bunched around 100 instead of ranging from 50 to 120.

Estimated rating 50

We see that the first observed time was 50 seconds, but the estimated rating of the worker was only 50, so he was only working at ***half*** the motivated or standard performance. Thus a motivated worker ***should*** be able to do the job twice as quickly, and the corrected or ***basic*** time is 50% of 50 seconds giving the figure of ***25 seconds.***

Estimated rating 90

Similarly the second basic time is 90% of 30 seconds or ***27 seconds.***

Estimated rating 120

In the third case, the operator was working faster than standard performance, so the basic time must be ***more*** than the observed time, i.e. 120% of 20 or ***24 seconds.***

Estimated rating 100

The fourth estimated rating was 100, so the basic time is the same as the observed time of ***26 seconds.*** The average of these four basic times for the element is, of course, ***25.5 seconds.***

Relaxation allowance

However, the basic time for an element is not the same as the standard time. We have to allow for the fact that the operator is a human being and cannot be expected to work steadily through the day like a machine. He will need to rest occasionally and, as the day wears on, he will get tired and his productivity will drop. To allow for this, we must make an addition, called the ***relaxation allowance***, to the basic time before we can calculate the overall standard time for the operation.

Percentage addition

The relaxation allowance is expressed as a percentage addition to the basic time, and naturally varies according to the type of job. There are tables of relaxation allowances available covering the various factors involved including:

Posture — whether bending and stretching is needed, as for digging a trench

Weight or force — depending on weight lifted or force exerted.

Concentration

Noise

Monotony
Fatigue
etc., etc.

Judgement

Although these tables of generally agreed relaxation allowances are available, good judgement is needed to use them properly. Some allowances, such as those for the force exerted to carry out the work, are directly measurable. But others, such as the fatigue experienced by a worker doing a particular job, can only be estimated by the work study officer on the basis of past experience.

Standard time for the whole job

After the relaxation allowance additions have been made for each element, the standard times for the separate elements are added (together with a contingency allowance) to give the ***standard time*** for the whole job. Since the work study officer has corrected his observations of basic time to the equivalent of '100' rating, the standard time for the whole job is the time in which the job should be completed by a competent operator working at ***standard performance.*** Allowance has been made for excessive strain and fatigue in carrying out the job, and the standard time can therefore be used to calculate the targets for an incentive bonus scheme to ensure that the workers are properly motivated.

Uses of time study

Although time study is a necessity if an incentive bonus scheme is to be scientifically-based, the results of work measurement techniques can be applied more generally. Time study provides the facts about human activity, and the contractor should aim to build up a library of facts and information covering the many different tasks that go into constructing a building. If this is done, he will be able to plan and control his operations much more effectively and it should be possible to take much of the 'guesswork' out of preparing estimates and tenders.

BUILDING BOOKS FROM INTERMEDIATE TECHNOLOGY PUBLICATIONS

Accounting & Book-keeping for the Small Building Contractor
Financial Planning for the Small Building Contractor
The Small Building Contractor and the Client

by Derek Miles

These three volumes are designed to help the small building contractor improve his accounting, planning and client-relationship methods. The subjects covered include: organising the office; basic book-keeping; analysis sheets; profit and loss; planning the year's work; job programmes; cash flow; investment decisions; billing procedures; work study techniques; estimating and tendering. Practical exercises and specimen forms are included.

Accounting & Book-keeping for the Small Building Contractor 190 pages. Illustrated. ISBN 0 903031 54 X. £3.95 net.
Financial Planning for the Small Building Contractor. Approx. 188 pages. Illustrated. ISBN 0 903031 55 8. £3.95 net.
The Small Building Contractor and the Client. In Preparation.

A Manual on Building Maintenance
Vol. 1: Management; Vol. 2: Methods

by Derek Miles

These companion volumes have been produced because of a lack of suitable guidance for managers of building units. This applies to both the public and private sectors in the administration and management of building sites, and in the technical aspects of maintenance work. Owing partly to the differing systems for allocating funds between capital and revenue budgets, expensive buildings are often allowed to deteriorate. This waste of resources is particularly unfortunate because repair and maintenace work is labour intensive and costly.

Vol. 1: Management deals with efficient control procedures such as resource budgets, finance, manpower, materials, equipment and provides a rational and practical system for measuring performance.

Vol. 2: Methods examines actual maintenance problems, suggests some of the more common causes of failure and sets out methods for dealing with them. These volumes make a useful and valuable contribution to increasing the operating efficiency of building units in developing countries and in the industrialised world.

> "Both volumes have been written to assist maintenance staff working in developing countries to appreciate and apply maintenance techniques . . . to their particular problems. In this, the books appear successful. In the first volume, methods of maintenance management and control are clearly set out and the book would be of great use to small maintenance departments in this country. Volume 2 . . . is clearly and simply written and illustrated and should be ideal for its intended use."
>
> — *Building Technology and Management,* June 1977

> "The manual will be of direct assistance to those struggling with the practical application of maintenance policies and procedures."
>
> — *Indian Concrete Journal,* August 1977

Vol. 1: Management 61 pages. Illustrated. ISBN 0 903031 28 0. £1.75 net.
Vol. 2: Methods 61 pages. Illustrated. ISBN 0 903031 40 X. £1.75 net.

Manual of Building Construction
by H.K. Dancy

A practical illustrated book on the construction of small buildings using local materials, suitable for a great variety of ground and climatic conditions. New edition, including metric conversion tables.
352 pages. Illustrated. ISBN 0 903031 08 6. £2.95 net.

> "A practical guide to anyone intending to produce a small house or group of buildings . . . It covers the whole range of operations from choosing the site, designing and laying out the building and each stage of construction. Means of using self-contained services for the treatment of sewage and generation of electricity are also discussed for operations in which publicly provided services are not available. Since it makes no assumptions regarding the previous training of the reader the work offers a valuable support for the novice builder."
>
> — *Building Design* (UK)

> "This manual contains a wealth of information on building from scratch."
>
> — *Rain* (USA)

Prices are correct at the time of going to press. The net price does not include postage and packing. Please add 15% to the price of publication for surface mail and UK; 35% for airmail. Orders should be sent to Intermediate Technology Publications Ltd, 9 King Street, London WC2E 8HN, UK, or to our overseas distributors. Details of these and a complete publications list are available on request.